Mathematics in Mind

SpringerBriefs in Cognitive Mathematics

SpringerBriefs present concise summaries of cutting-edge research and practical applications across a wide spectrum of fields. Featuring compact volumes of 50 to 125 pages, the series covers a range of content from professional to academic. Briefs are characterized by fast, global electronic dissemination, standard publishing contracts, standardized manuscript preparation and formatting guidelines, and expedited production schedules.

Typical topics might include:

- A timely report of state-of-the art techniques
- A bridge between new research results, as published in journal articles, and a contextual literature review
- A snapshot of a hot or emerging topic
- An in-depth case study
- A presentation of core concepts that students must understand in order to make independent contributions.

SpringerBriefs in Cognitive Mathematics showcases research in the study of math cognition as well as the relation of math to other faculties, such as language, music, and art, among others. As a subseries of **Mathematics in Mind**, the research published in this series falls under the aegis of the Fields Cognitive Science Network, which brings together mathematicians and cognitive scientists to explore the nature of mathematics from various interdisciplinary angles. All works are peer-reviewed to the highest standards of scientific literature.

Josip Slisko

Fostering Cognitive Mathematics Skills with Matchstick Puzzles

A Guide for Researchers, Teachers, and General Readers

Springer

Josip Slisko
Facultad de Ciencias Físico Matemáticas
Benemérita Universidad Autónoma de
Puebla
Puebla, Mexico

ISSN 2522-5405 ISSN 2522-5413 (electronic)
Mathematics in Mind
ISSN 3091-4213 ISSN 3091-4221 (electronic)
SpringerBriefs in Cognitive Mathematics
ISBN 978-3-032-14560-4 ISBN 978-3-032-14561-1 (eBook)
https://doi.org/10.1007/978-3-032-14561-1

This Springer imprint is published by the registered company Springer Nature Switzerland AG
The registered company address is: Gewerbestrasse 11, 6330 Cham, Switzerland

The book is dedicated to my wife Jesenka, and my son Javor. Without their understanding and support, the book wouldn't have been written.

Preface

The purpose of the book is to critically analyze the past and actual usage of matchstick puzzles *with multiple solutions.*

I strongly believe that these puzzles, if properly used, have a great potential to increase cognitive mathematics skills of puzzle solvers of every age. These skills are:

(1) strategic analysis and solution planning;
(2) mathematical creativity;
(3) critical thinking;
(4) avoiding self-imposed cognitive biases;
(5) visual intelligence; and
(6) systematic combinatory thinking.

Unfortunately, as this book shows by various examples, in too many multiple solution puzzles only one solution is asked or given to puzzle solvers. Being so, *the chances for increasing cognitive mathematics skills are drastically reduced or completely lost.*

In addition, when a reader finds a correct solution which is different from the given in the book's end, she or he might think that her or his solution is wrong. If this situation repeats with a few puzzles, the reader might create an *undeserved negative feeling about her or his creative abilities.* This situation opens an issue of ethical responsibility of authors and publishing companies toward the readers who spend money to buy their books!

The book also calls attention to a controversial theoretical issue: which solutions of a matchstick puzzle should be considered different? Namely, it seems that many authors, rather implicitly than explicitly, consider that additional (visually different) solutions than can be obtained by rotation or mirror reflection from a solution are not essentially different ones and, consequently, they neither publish nor mention them.

The book presents promising experimental results, obtained in few pilot explorations with different-age student and pupil groups, showing that many puzzle solvers, if informed about the number of possible visually or fundamentally different solutions of a puzzle, become able to find the solutions that were never published in the books, even in those written by recognized authors.

The book is, as far as I know, the first and long-time needed extensive critical review of the multi-faceted phenomenon of matchstick puzzles uses in different domains, from pure entertainment activity and puzzle-based mathematics learning to psychological research on problem solving and prevention of cognitive decline.

The main benefit for readers would be to recognize, based on presented documental and experimental evidence, a great potential that matchstick puzzles with multiple solutions have for fostering cognitive mathematics skills and self-confidence of puzzle solvers.

It is known that the main obstacle to have more creative and innovation thinking in educational and other social domains are false beliefs (held by too many persons due to traditional mathematics education):

Every problem has only one solution and only one solving method.
Only few exceptional persons have innate skills to think creatively.

The book provides many examples of puzzles with multiple solutions and examples of uncomplete solutions by famous authors that show the falsehood both beliefs. In addition, the persons involved in a few experimental explorations were able to find the solutions missed by those authors. What a great feeling is being able to find solutions that were not imagined by famous book authors!

In addition, other cognitive biases of authors, like "all sought squares or triangles should be equal" or "formulation change doesn't matter" are presented, too.

The book also offers many examples of analyses of finding missing solutions in multiple-solution matchstick puzzles. These examples might help readers develop own cognitive mathematics skills or cognitive mathematics skills of other persons they care about (children, pupils, students, patients...).

I hope this book will be useful to various groups of readers.

Researchers in cognitive psychology and educational mathematics, interested in using matchstick puzzles to explore experimentally the relationships between different cognitive skills and levels of performance in solving matchstick puzzles, can find in the book various examples of detailed description of strategic analysis and planning of possible solutions of matchstick puzzles.

Mathematics teachers, at different educational levels, can find in the book description and theoretical justification of didactical design elements for using matchstick puzzles to foster development of cognitive mathematics skills and puzzle posing abilities of their students.

General readers might improve their own cognitive mathematics skills. Parents, who want to develop the same skills in their children, can help them avoid common false beliefs ("every problem has only one solution and one solving methods" and "only exceptional persons have an innate skill for creative thinking"). Parent can also encourage children to explore their creative abilities in posing and solving own matchstick puzzles.

Finally, I acknowledge that I was greatly benefited by the frequent email communications with Heinrich Hemme from the University of Applied Sciences in Aachen (Germany). He kindly shared with me his expert knowledge on finding fundamentally different solutions to multiple-solution matchstick puzzles and an impressive unpublished collection of matchsticks puzzles and solutions. Both were very helpful in finding missing solutions reported and discussed in the book.

I would be very grateful to those readers who would kindly share with me those solutions of presented multi-solution matchstick puzzles that were unknown to me.

Puebla, Mexico

Josip Slisko
jslisko@fcfm.buap.mx

Contents

Chapter 1
Puzzle-Based Mathematics Learning

Abstract The aim of this chapter is to give a short summary of puzzle-based mathematics teaching and learning. Mathematical puzzles were used in mathematics teaching since long time ago in many countries. Due to curricular pressure, in a period mathematics textbook authors were avoiding using mathematical puzzles or used very few of them. Recently, this unfortunate situation started to change. Namely, many cognitive and motivational benefits of solving mathematical puzzle are being recognized and puzzle-based learning is increasing its presence in education.

1.1 Puzzle-Based Learning in the Past and Recent Changes

Riddles and puzzles have been part of human culture since its beginnings. Many think they serve only an entertaining function, like good jokes or songs suitable for brightening up leisure time. However, their great popularity and longevity demonstrate that they express a deeper, almost instinctive, human inclination toward mysteries and the unknown (Danesi 2002).

In addition to qualitative puzzles that challenge human conceptual ingenuity, there are quantitative puzzles that are much more than just fun and whose solution, in addition to a proper understanding of their essence, requires some mathematical knowledge. These mathematical puzzles are the fundamental basis of "recreational mathematics". There is an impressive bibliography of books that have been published in this area over the centuries (Schaaf 1955).

At the end of the eighth century, Alcuin wrote the first collection of tasks for the mathematical education of young people to "sharpen their mind". In the collection were included, along with practical problems, some quantitative and procedural puzzles with no immediate application in the real world (Hadley and Singmaster 1992).

Fibonacci wrote his book "*Liber abaci*" at the very beginning of the thirteenth century (Sigler 2002) for teaching commercial and practical mathematics with using Arabic numbers. The book additionally contained problems related to "puzzling situations" that hardly could happen (Sigler 2002, p. 261):

J. Slisko, *Fostering Cognitive Mathematics Skills with Matchstick Puzzles*, Mathematics in Mind, https://doi.org/10.1007/978-3-032-14561-1_1

> There are two men whose propose to go to a long journey, and one will go 20 miles daily. The other truly goes 1 mile the first day, 2 the second, 3 the third, and so on always one more mile daily to the end when they meet. It is sought for how many days the first is followed.

Puzzling "word problems" were also used in mathematics teaching in ancient times in China, India, and Islamic countries (Swetz 2012).

This initial symbiosis of mathematics teaching and mathematics puzzles almost ceased to exist today. Mathematics textbooks must expose the content suggested by curricular requirements, while recreational mathematics books are shaped by the creativity, and criteria of the authors who try to be in resonance with eventually interests of general readers. Being so, mathematics textbooks and recreational mathematics books gradually diverged to become parts of two separate publishing areas for different groups of readers and users.

Martin Gardner, an undisputed legend in the formulation and popularization of fun mathematical puzzles and games through a regular column in the prestigious journal "*Scientific American*", in the Introduction of his book "*Mathematical circus*" wrote the following words (Gardner 1979, p. XI):

> (Recreational mathematics)… is the best way to capture young people's interest in teaching elementary mathematics. A good mathematical puzzle, a paradox, or a magical trick can stimulate children's imagination much faster than a practical application (especially if the application is remote from the child's experience)…

Although Gardner strongly advocated for the inclusion of play and magic in mathematics education, he was not pleased with the results (Gardner 1998):

> The monthly magazine published by the National Council of Teachers of Mathematics, Mathematics Teacher, often carries articles on recreational topics. Most teachers, however; continue to ignore such material. For 40 years, I have done my best to convince educators that recreational mathematics should be integrated into the standard curriculum. It should be regularly presented as a way to interest young students in the wonders of mathematics. So far, though, movement in that direction has been glacial

Fortunately, the situation regarding the relationship between fun mathematics and puzzles in mathematics education has changed considerably and now looks much better than it did 20 or 30 years ago.

1.2 Puzzles in Mathematics Education

An increasing number of authors are suggesting the use of puzzles in mathematics teaching for different purposes and effects on student learning. For example, Jelle (2017) states the following:

> The use of puzzles in teaching can increase motivation, improve understanding, promote creative processes, expand the ability to engage and solve diverse challenges from various perspectives, and thus lead to improved learning processes and problem-solving skills.

Ben-Chaim and collaborators (2019) argue:

> The inclusion of mathematical puzzles, even in a small way, in the mathematics classroom produces considerable interest among students. Some even join extracurricular classes and become "addicted" to solving puzzles in mathematics and other areas... Puzzle-solving encourages original thinking, creativity, and intuitive reasoning that often spans different fields of study and the search for unconventional solution methods. Furthermore, and no less importantly, mathematical puzzles and riddles reveal the richness, beauty, and wisdom embodied in mathematics.

After generally defining recreational mathematics as "a type of play that is enjoyable and requires mathematical thinking and skills to engage with," Rowlett and collaborators (2019) state:

> (Recreational mathematics)... is accessible to a wide range of people and can be used effectively to motivate engagement and develop understanding of mathematical ideas or concepts. Recreational mathematics can be used in education for engagement and to develop mathematical skills, sustain interest during procedural practice, and challenge and stretch students. It can also make cross-curricular links, including with the history of mathematics.

Kurnik (2008) describes the characteristics of "fun tasks" that can help develop logical reasoning and ingenuity, and spark interest in mathematics:

> 1) The fun tasks are mathematical miniatures, the solution of which requires only the most basic knowledge of arithmetic, algebra, and geometry.
>
> 2) The task formulations are simple and understandable to everyone.
>
> 3) The texts are written in the form of clever little stories from everyday life.
>
> 4) Greater prior mathematical knowledge is not always a guarantee of faster resolution.
>
> 5) The problems are not always easy. Many of them require considerable mental effort, logical reasoning, and especially ingenuity to find a way to solve them.

Mental efforts needed to solve mathematical puzzles are related to the following fact: a good puzzle commonly causes a "cognitive bias", leading to an erroneous answer. The concept of "cognitive bias" and its important implications are today widely recognized and described in different domains:

> Cognitive biases are errors in memory or judgment that are caused by the inappropriate use of cognitive processes. These biases are caused by the overuse of schemas, the reliance on salient and cognitive accessible information, and the use of rule-of-thumb strategies known as heuristics… Understanding the potential cognitive errors we frequently make can help us make better decisions and engage in more appropriate behaviors. (Stango 2013, p. 256)
>
> ... Cognitive biases are systematic, unconscious, automatic patterns of thought that sometimes distort thinking and potentially lead to errors. (Howard 2018, p. 14)
>
> Cognitive biases are mistakes in reasoning or other cognitive processes, produced by information-processing rules that the brain uses to make decisions. (Whitesmith 2020, p. 2)
>
> Cognitive biases are systematic patterns of deviation from rational judgment, representing how human cognition leads us astray in decision-making. These biases emerge from the brain's information processing, often resulting from mental shortcuts known as heuristics. Heuristics help the brain simplify complex problems by providing quick, intuitive judgments, but they can also lead to errors in reasoning. (Lemieux 2025, p. 14)

1.3 Puzzles in Engineering Education and Economics

Individual quantitative, visual, logical, or manipulative puzzles have long been used as useful tasks in the study of cognitive processes (Allport 1997) and in some textbooks teaching general strategies for improving thinking, learning, and creativity (Bransford and Stein 1993).

However, books and articles on how to systematically use puzzles to learn critical thinking and problem-solving skills have only recently begun to appear in engineering education (Michalewicz and Michalewicz 2008; Thomas et al. 2013; Meyer et al. 2014).

Puzzle-based learning is considered effective in the systematic and timely development of such skills for the following reasons (Michalewicz and Michalewicz 2008, p. XII):

> Puzzles are instructive because they illustrate useful (and powerful) rules for problem solving in a very fun way. Puzzles are fascinating and stimulating. Contrary to many textbook problems, puzzles are not "attached" to any chapter (as is the case with real-world problems).
>
> It is possible to talk about different techniques (e.g., simulation and optimization), disciplines (such as probability and statistics), or application areas (e.g., programming and finance) and illustrate their significance by discussing a few simple puzzles. At the same time, students are made aware that many of the conclusions are applicable in the broader context of real-world problem solving.

It's important to emphasize that in recent years, Microsoft, Google, and other high-tech companies have used large number of logical and mathematical puzzles and "impossible questions" in their extremely challenging job interviews (Poundstone 2003, 2012; Kador 2005). "Puzzle interviews" have become a new trend. From Wall Street to Silicon Valley, employers use difficult questions to assess a candidate's intelligence, imagination, and problem-solving ability. These are essential skills for survival and success in today's competitive global marketplace.

Managers seeking the most talented employees need to learn how to incorporate good (and unfamiliar) puzzles into the conversations they will lead in their search for the best candidates. Job seekers must discover how to deal with "brain-exploding" problems and how to gain an edge that could lead to a job for life. John Kador, a renowned expert on job interview questions, describes well the economic reasons for this strategy (Kador 2005, p. VI):

> The use of puzzles and riddles makes sense for companies that focus their recruiting efforts more on what candidates can do in the future than on what they have done in the past. These companies understand that in today's fast-paced global business world, specific skills are of limited use because technology changes so rapidly. What is really needed, according to interviewers, are curious, observant, and resourceful candidates who embrace new challenges, demonstrate mental agility under stressful conditions, learn quickly, defend their thinking, and demonstrate enthusiasm for impossible tasks.

It is worth noting that the "*Cognitive Reflection Test*" (Frederick 2005), frequently used to identify "fast thinkers" and "slow thinkers" (Kahneman 2012) among economics students, consists of three well-known mathematical puzzles. The third puzzle reads:

> In a lake, there is a patch of lily pads. Every day, the patch doubles in size. If it takes 48 days for the patch to cover the entire lake, how long would it take for the patch to cover half of the lake? _____ days

"Fast thinkers" give the incorrect answer "24 days". Their "cognitive bias", leading to this incorrect answer, is to suppose that the relationship between the time and the area of lake covering is linear one. This "cognitive bias", named "illusion of linearity", is caused by common practice in mathematics education. Students are given mainly problems that are correctly solved by the rule of three (Van Dooren et al. 2004).

"Slow thinkers", through "cognitive reflection", can escape "cognitive bias", critically analyzing the implication of incorrect answer: If half of the lake is covered in 24 days, and every day, the patch doubles in size, then the whole lake would be covered not in 48 but in 25 days!

Such a careful logical analysis rejects the answer "24 days" and leads easily to correct answer "47 days".

Some research results indicates that "fast" thinkers (those who give three incorrect answers on the test) are the ones who make mistakes more frequently on economic tasks that require accurate information processing and decision-making (Toplak et al. 2011).

All these educational trends and economic developments require an increased use of mathematical puzzles in mathematics teaching through the development, implementation, and evaluation of new designs for their productive didactic uses. The essence of these designs is related to learning processes that make possible that puzzle solvers became able to escape related cognitive biases.

References

Allport, G. W. A. (1997). Planning and problem solving using the five-disc Tower of London task. *The Quarterly Journal of Experimental Psychology: Section A, 50*(1), 49–78.

Ben-Chaim, D., Shalitin, Y., & Stupel, M. (2019). Historical mathematical problems suitable for classroom activities. *The Mathematical Gazette, 103*(556), 12–19.

Bransford, J. D. & Stein, B. S. (1993). *The Ideal Problem Solver. A Guide for Improving Thinking, Learning, and Creativity*. Second Edition. W. H. Freeman and Company.

Britton, B., & Tayeh, C. (2004). Egg dilemma. *Teaching Children Mathematics, 11*(1), 26–28.

Danesi, M. (2002). *The Puzzle Instinct: The meaning of puzzles in human life*. Indiana University Press.

Frederick, S. (2005). Cognitive reflection and decision making. *Journal of Economic Perspectives, 19*(4), 25–42.

Gardner, M. (1979). *Mathematical circus*. Alfred A. Knopf.

Gardner, M. (1998). A Quarter-Century of Recreational Mathematics. *Scientific American, 279*(2), 68–75.

Hadley, J. & Singmaster, D. (1992). Problems to Sharpen the Young. *The Mathematical Gazette, 76*(475), 102–126.

Howard, J. (2018). *Cognitive Errors and Diagnostic Mistakes: A Case-Based Guide to Critical Thinking in Medicine*. Springer.

Jelle, B. P. (2017). Reviewing the Learning Process through Creative Puzzle Solving. *Creative Education, 8*, 2009–2035.

Kador, J. (2005). *How to ace a brainteaser interview*. McGraw-Hill.

Kahneman, D. (2012). *Pensar rápido, pensar despacio*. Barcelona: Debate.

Kurnik, Z. (2008). Zabavna matematika. *Matematika i škola, 45*, 196–202 (published in Croate language).

Lemieux, F. (2025). *Intelligence Failures and Strategic Surprises in Complex Geopolitical Environments*. Emerald Group Publishing.

Meyer, E. F., Falkner, N., Sooriamurthi, R., & Michalewicz, Z. (2014). *Guide to Teaching Puzzle-based Learning*. London: Springer.

Michalewicz Z, Michalewicz M (2008) P*uzzle-based learning: an introduction to critical thinking, mathematics, and problem solving*. Melbourne: Hybrid Publishers.

Poundstone, W. (2003). *How Would You Move Mount Fuji? Microsoft's Cult of the Puzzle. How the World's Smartest Companies Select the Most Creative Thinkers*. Little, Brown and Company.

Poundstone, W. (2012). *Are you smart enough to work at Google? Fiendish Puzzles and Impossible Interview Questions from the World's Top Companies*. Oneworld Publications.

Rowlett, P., Smith, E., Corner, A. S., O'Sullivan, D., & Waldock, J. (2019). The potential of recreational mathematics to support the development of mathematical learning. *International Journal of Mathematical Education in Science and Technology, 50*(7), 972–986.

Schaaf, W. L. (1955). *A Bibliography of Recreational Mathematics*, Volumes I–IV, National Council of Teachers of Mathematics.

Sigler, L. E. (2002). *Fibonacci's Liber abaci: a translation into modern English of Leonardo Pisano's Book of calculation*. Springer.

Stango, C. (2013). *Introduction to Psychology*. Flat World Knowledge.

Swetz, F. J. (2012). *Mathematical expeditions. Exploring word problems across the ages*. The Johns Hopkins University Press.

Thomas, C., Badger, M., Ventura-Medina, E., & Sangwin, C. (2013). Puzzle-based learning of mathematics in engineering. *Engineering Education, 8*(1), 122–134.

Toplak, M. E., West, R. F., & Stanovich, K. E. (2011). The Cognitive Reflection Test as a predictor of performance on heuristics-and-biases tasks. *Memory & cognition, 39*(7), 1275–1289.

Van Dooren, W., De Bock, D., Hessels, A., Janssens, D., & Verschaffel, L. (2004). Remedying secondary school students' illusion of linearity: A teaching experiment aiming at conceptual change. *Learning and instruction, 14*(5), 485–501.

Whitesmith, M. (2020). *Cognitive bias in intelligence analysis: Testing the analysis of competing hypotheses method*. Edinburgh University Press.

Chapter 2
Matchstick Puzzles: Early History and a Critical Review of a Few Recent Books

Abstract The first part of this chapter presents a summary of early history of the phenomenon of matchstick puzzles, from its beginning in the Year 1850 to the appearance of ground-breaking book Tromholt "Matchstick games. Mental activity and entertainment", published in the Year 1889 (Tromholt 1889). In the beginning geometric matchstick puzzles were published as single entertainment pastime in diverse journals and newspapers. In the Year 1865 German school teacher Werner published the first article with various matchstick puzzles proposing them as learning activities in home schooling. This idea was overtaken very soon by book authors like Ferdinand Braun (Der junge Mathematiker und Naturforscher. Verlag von Otto Spamer, 1876) and Louis Mittenzwey (Mathematische Kurzweil. Julius Klinkhardt, 1880) whose books prepared the stage for the book of Tromholt. Important contributions of Tromholt's book to the world of matchstick puzzles will be analyzed in Chap. 5. The second part of this chapter describes the results of a critical review of a few recent books on matchstick puzzles.

2.1 Popularity of Matchstick Puzzles

Matchstick puzzles are thinking problems in which the tasks are to transform initial matchstick configuration into a sought matchstick configuration by a specific procedure that consist of removing, moving, or adding a particular number of matchsticks. They are very popular today, both on many Internet pages and on the YouTube channel, as well as in a significant number of printed or Kindle books. They can be formulated and solved using classic wooden matchsticks, toothpicks, ear sticks, or even paper strips. Of course, today there is also a possibility of using "virtual matches" in interactive versions of these puzzles.

Unlike crosswords or sudoku games, matchstick puzzles no longer appear in the columns of daily newspapers. There isn't a serious specialized journal which publishes matchstick puzzles with the aim to refine their goals, form, and content, through collective efforts of the authors and readers. However, the good news is that these puzzles have won their presence in various fields of learning and research as

J. Slisko, *Fostering Cognitive Mathematics Skills with Matchstick Puzzles*, Mathematics in Mind, https://doi.org/10.1007/978-3-032-14561-1_2

well as in mathematics textbooks, collections of fun mathematics, and books written for practitioners for related or seemingly unrelated professions. Nevertheless, in some cases the authors reveal various cognitive biases. These biases will be described and commented in the Chap. 3.

For current and potential users of matchstick puzzles in school or extracurricular forms of education, it might be informative to know something about their early history and some unfortunate features of recent books on matchstick puzzles.

2.2 Early History: Matchstick Puzzles in Magazines

The first puzzle, which is the forerunner of today's geometric matchstick puzzles, was published in 1850 in the second volume of a collection of articles in the highly successful English family magazine "*The Family Friend*" (Fig. 2.1).

The "*Family Pastime*" section, on page 148, contained the following "practical puzzle" (Fig. 2.2).

Fig. 2.1 Cover of the second volume of a collection of articles published in the magazine "The Family Friend" (*Source* "The Family Friend", 1850)

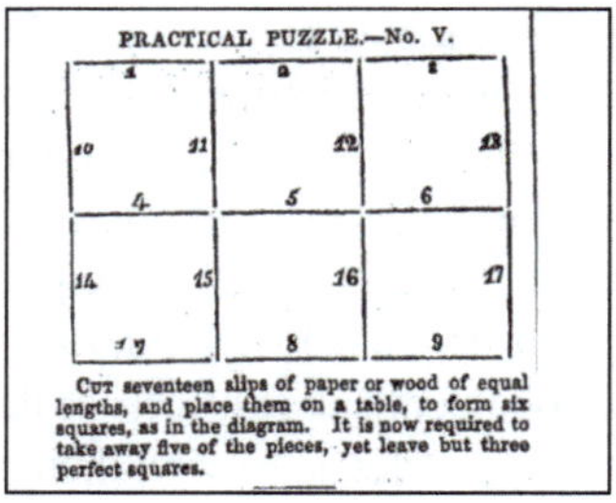
PRACTICAL PUZZLE.—No. V.

Cut seventeen slips of paper or wood of equal lengths, and place them on a table, to form six squares, as in the diagram. It is now required to take away five of the pieces, yet leave but three perfect squares.

Fig. 2.2 The precursor to today's geometric matchstick puzzles (*Source* "The Family Friend", 1850)

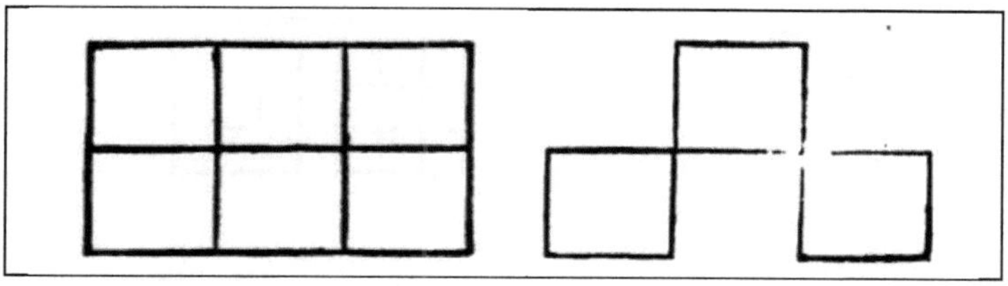

Fig. 2.3 Initial and final configuration of lucifer matches in Arnold's puzzle "The magic square" (*Source* Arnold 1858, p. 286, p. 301)

The puzzle text was: "Cut seventeen pieces of paper or wood of equal length and place them on the table to form six squares as shown. It is now required to take away five of the pieces, yet leave but three perfect squares remain".

The solution, published on page 179 of the collection, was verbal: "Remove the pieces marked with the numbers 8, 10, 1, 3, and 13 and only three squares will remain". The puzzle and solution were repeated in the same form in several books over the following years, with two significant shortcomings: no drawing of the solution and no mention of another possible, visually different solution (remove the pieces marked 2, 14, 17, 9, and 17).

The first flaw was corrected in the Year 1868, when magician George Arnold published the puzzle "*The magic square*" with initial and final configurations of lucifer matches (Fig. 2.3).

The second flaw, unfortunately, is still present today: the second solution, which a mirror image or rotated version of the first solution, is not mentioned in either magazines or books!

This is a doubly strange fact today. On the one hand, the authors of the first books using matchstick puzzles began to mention or directly show drawings of multiple solutions.

On the other hand, the authors' silence about the possibility of different solutions promotes and perpetuates the erroneous belief that every mathematical problem has only one solution. Such a belief hinders the development of creative thinking, which is one of the most important skills for a successful life in the twenty-first century. The intellectual beauty of some famous matchstick puzzles consisted precisely in the fact that there were several possible solutions that were later discovered by particularly creative readers, and that the authors of the original puzzles did not notice the existence of those solutions when they formulated and published them! A few cases will be described in Chap. 9.

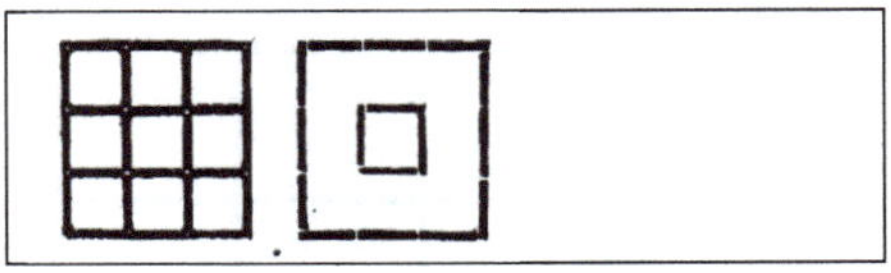

Fig. 2.4 Remove 8 matches and form 2 non-touching squares (*Source* Werner 1865)

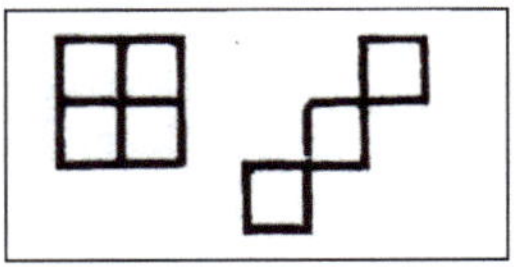

Fig. 2.5 Move 4 matches and form 3 squares from 4 squares (*Source* Werner 1865)

The next important episode in the early history of matchstick puzzles was the publication of magazine articles in which not only one puzzle appeared, but several puzzles were presented to the readers. An Internet search on www.archive.org of the available literature indicates that this (probably) first happened in 1865 in the German journal "*Cornelia*", which was intended as a "home education magazine". In his article, the schoolteacher Werner (personal name not mentioned!) presented several original geometric puzzles (Werner 1865). Examples are a "match removal puzzle" (Fig. 2.4) and a "match moving puzzle" (Fig. 2.5).

Both puzzles appeared a quarter of a century later in the first American magazine "collection" consisting of only three toothpick puzzles. It was first published in the magazine "*The New York Herald*" on January 25, 1891. The "collection" was repeated the same year, almost verbatim, by "*The Boston Weekly Globe*" on February 10 (with a mention of the original publication) and "*The Polk County Journal*" on September 3 (without mentioning the magazine from which it was "borrowed").

Werner was also the first author to connect matchstick configurations with Roman numerals, "proving" that half of the number twelve can be equal to seven! For people whose thinking is "framed" by Arabic numerals, such a claim is absurd. However, if 8 matchsticks are used to make the Roman numeral XII (Fig. 2.6a), then its "upper half" is the Roman numeral VII made of 4 matchsticks (Fig. 2.6b).

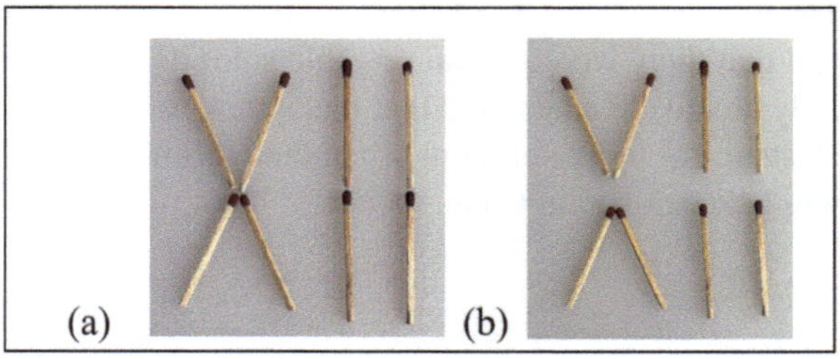

Fig. 2.6 "Proof" that half of the number twelve is equal to the number seven (*Source* Author)

2.3 Early History: Matchstick Puzzles in Books

The first book to include ten different geometric matchstick puzzles was published in 1876 in Germany. It was written by Ferdinand Braun, future Nobel-prize winner in physics (1909). At the time of writing, Braun was young physics and mathematics teacher in Leipzig. The book had the title "*Young mathematician and natural scientist*" (Braun 1876).

Being so, Braun recognized the potential of matchstick puzzles solving to help young people develop thinking skills necessary for learning and doing mathematics and physics. The presentation of included matchstick puzzles was a pedagogically structured dialog between a fictious father and his six children (Adolf, Anna, Fritz, Gustav, Max, and Otto).

In a contextualized version of the first matchstick puzzle (Fig. 2.3), Braun "predicted" a likely children's error in its solving: presence of a single, free matchstick in the puzzle's "solution".

Braun was the first book author who presented more than one solution to a puzzle. In the first case, the formulation, initial matchstick configuration, and two solutions found by children (Gustav and Adolf) are given in Fig. 2.7.

The most important Braun's contribution was to present a numerical strategy for finding structural features of sought matchstick configuration. In posterior chapters it will be called "Braun's strategy". The first example was given with following puzzle (Fig. 2.8).

Braun's numerical strategy was: "If you count all matches in that figure, … you will find that there are 17 of them… If… 6 shall be taken away and still 3 squares shall remain, one square will have to be free and two will have to touch at an edge". This conclusion was obtained by the simple calculation: "11 edges shall remain. 1 free square has 4 edges. 2 squares touching at an edge have 7 edges. In sum, there are 11 edges".

Braun found three different solutions with this matchstick configuration (Fig. 2.9).

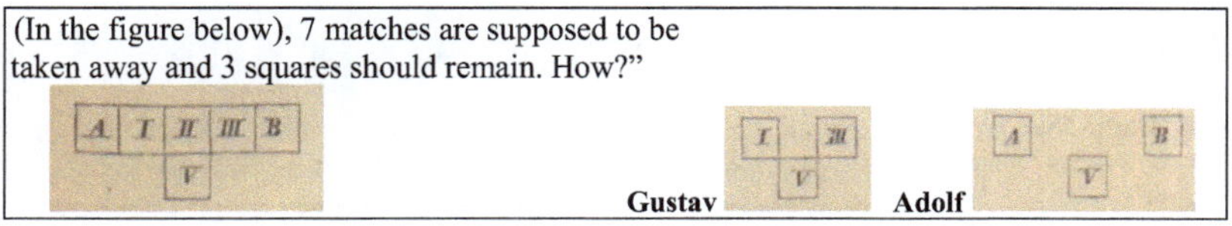

Fig. 2.7 The formulation, initial matchstick configuration two solutions to a multiple-solution puzzle (*Source* Braun 1876)

Fig. 2.8 The formulation and initial matchstick configuration for which numerical strategy was presented (*Source* Braun 1876)

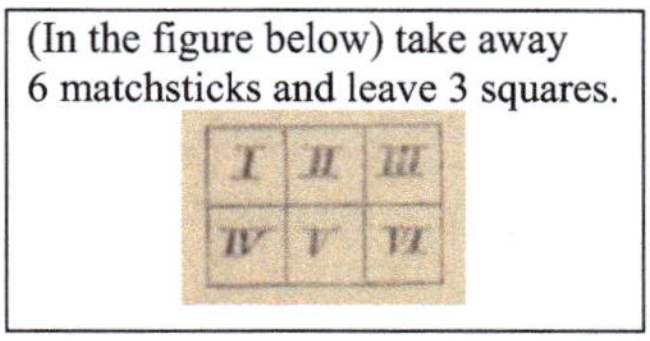

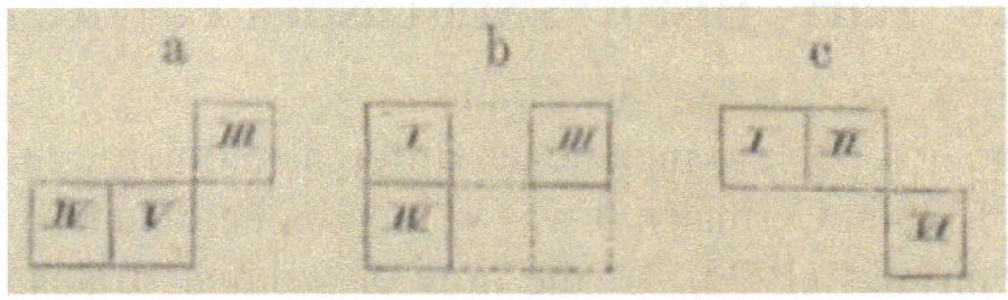

Fig. 2.9 Three different solutions to the puzzle “Take away 6 matchsticks to get 3 squares” (*Source* Braun 1876)

After learning that strategy, Gustav was able to find two additional solutions to the previous puzzle (Fig. 2.10).

Having big pedagogical intuition, Braun found important to describe two possible outcomes of using his numerical strategy by children. It was done with the following matchstick puzzles (Fig. 2.11).

The first outcome is consequence of predicted cognitive bias “remaining square must be equal”. Otto, who suffered that bias, said:

> There are 12 matches in that figure, take away 2 and 10 remain. 10 edges are always too many for 2 squares, however, whence the *problem is erroneous.*

Believer in the positive outcome of group learning discussion, Braun “gave” to Max the following closing words:

> For two small squares there would certainly be too many edges, so there must be a bigger square among them. The smallest which can be laid out with matches of equal length, however, has 8 matches if 4 single matches are insufficient. Accordingly, there are only 2 edges left for the second square. So, it must share 2 edges with the big square, ergo the solution is as in Fig. 2.12.

The transfer of the numerical strategy to a new matchstick configuration and task was done with the following puzzle (Fig. 2.13).

Braun gave words to Otto to show the transfer of numerical strategy:

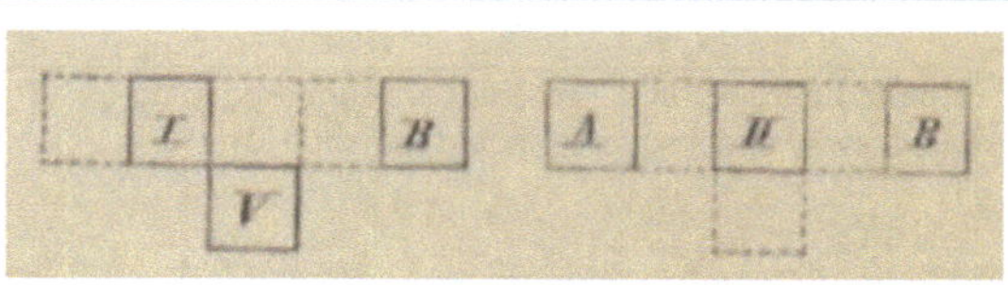

Fig. 2.10 Two additional Gustav’s solution to the puzzle “Take away 7 matchsticks to get 3 squares” (*Source* Braun 1876)

You are supposed to take away two matches so that 2 squares remained. Solve that task according to the rule above, systematically that is, without trying around.”

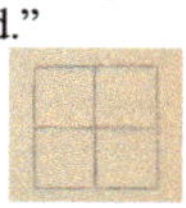

Fig. 2.11 The puzzle for using numerical strategy (“rule” in Braun’s word) in solving it (*Source* Braun 1876)

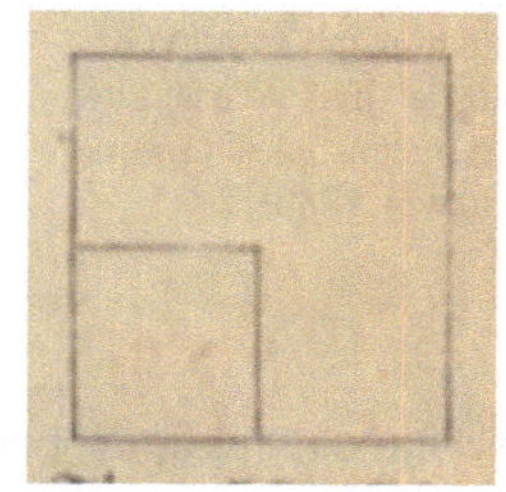

Fig. 2.12 Max's solution to the puzzle by using correctly the numerical strategy ("rule" in Braun's word) in solving it (*Source* Braun 1876)

(In the figure below), which consists entirely of equilateral triangles,
5 edges shall be taken away and 5 triangles shall remain. How can you approach this?

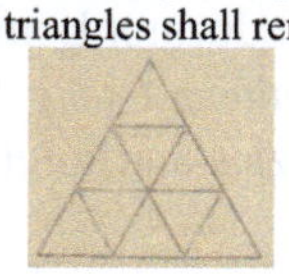

Fig. 2.13 The puzzle to illustrate the transfer of the numerical strategy to a new matchstick configuration and task (*Source* Braun 1876)

> First, I count the number of edges – 18. If I take away 5 of them, 13 will remain. The task therefore is, to make 5 triangles from 13 edges.

As for 5 triangles with no shared sides 15 matchsticks are needed, Braun considered different possibilities for side sharing. For two triangles with one shared side, 5 matchsticks (or "edges" in Braun's terminology) are needed. For three triangles with two shared sides, 7 matchsticks are needed. Being so, there are two possible configurations for making 5 triangles from 13 matchsticks:

1. two triangles with no shared sides (6 matchsticks) and three triangles with 2 shared sides (7 matchsticks);
2. one triangle with no shared sides (3 matchsticks), two triangles with one shared side (5 matchsticks), and two triangles with one shared side (5 matchsticks).

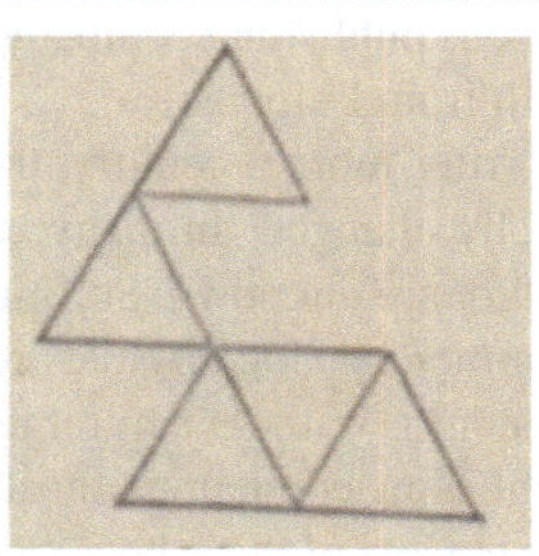

Fig. 2.14 The solution with two triangles with no shared side and three triangles with two shared sides (*Source* Braun 1876)

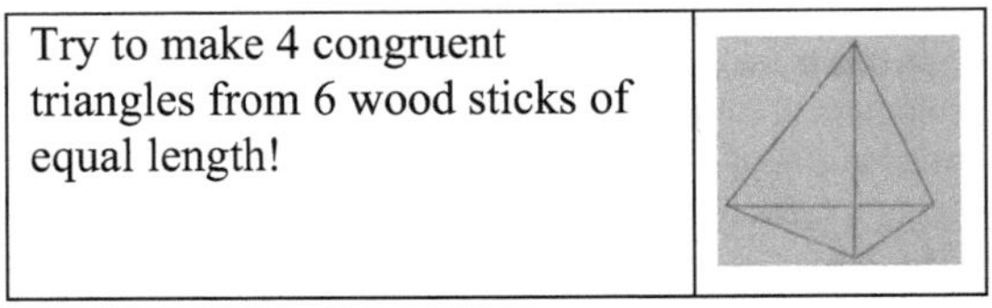

Fig. 2.15 Mittenzwey's formulation and solution to the puzzle "make 4 triangles from 6 matchsticks" (*Source* Mittenzwey 1880)

Nevertheless, Braun published only one solution with the first configuration (Fig. 2.14).

It was suggested to the readers to find other possible solutions, without indication about how many they were.

The next book with 27 matchstick puzzles was published in the Year 1880 by mathematics teacher and prominent educator Louis Mittenzwey from Leipzig. It had the title "*Mathematical Fun*" ("*Mathematische Kurzweil*"). Although Mittenzwey "borrowed" various matchstick puzzles and solutions from Braun's book, he didn't give the credit to Braun.

For Werner's puzzle "remove 8 matchsticks to get 2 non-touching squares" (Fig. 2.4), Mittenzwey only published Werner's solution (Fig. 2.4) although in formulation he omitted the crucial restriction that remaining squares should be "non-touching". As for "touching" squares additional solutions are possible, that was the first case of cognitive bias "formulation change doesn't matter". The presence of this cognitive bias among many book authors will be presented in Chap. 7.

In his book, Mittenzwey published the puzzle that today is considered a paradigmatic example of problems whose solution requires "thinking-out-of-box" (Mittenzwey 1880, p. 32, p. 84). Its formulation and solution are given in Fig. 2.15.

2.4 A Short Critical Review of Some Recent Books on Matchstick Puzzles

At the beginning, from 1876 to 1889, books on matchsticks puzzle were written by recognized professionals with mathematical knowledge, like Ferdinand Braun, Louis Mittenzwey, and Sophus Tromholt (whose big contribution to the world of matchstick puzzle will be presented in Chap. 5). These books were published by serious publishing companies, like *Otto Spammer Verlag* or *Julius Klinkhardt Verlag*, both from Leipzig.

After more a century, the situation with books on matchstick puzzles has drastically changed. In many cases, name of publishing company or/and name of the author are unknown. Self-publishing makes possible to create books, with appealing commercial slogans, that are sold with the help of Amazon (print-on-demand selling strategy). Without a minimum quality control, commonly carried out in professional publishing companies, those books contain different kinds of serious errors. Being so, the readers of these books not only wasted their money but would be likely confused with some "solutions".

To illustrate this unfortunate situation, here come a few alarming examples.

Move 3 matchsticks to get 4 squares.

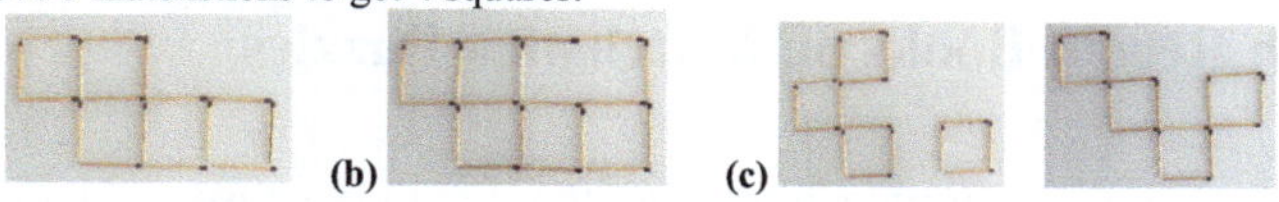

(a) (b) (c)

Fig. 2.16 The Puzzle 56 without relationship between the formulation and the published "solution" (*Source* Author)

Move four matchsticks to create three squares.

Fig. 2.17 The example of a mindless error in the published "solution" to a geometric matchstick puzzle (*Source* Author)

The book "*Math puzzles matchstick*" allegedly offers: "**expanded gameplay**... that will keep you engaged and eager to discover new strategies", "**diverse equations** ... that sharpen your mathematical skills", "**enhanced strategic thinking** ... brings you closer to unraveling the mathematical mysteries", and "**verify** your answers as you progress through extended gameplay".

The ironical truth is that the books don't give to the readers a single puzzle solution to verify their answers!

The book "*100 matchstick puzzles. Fun family activities*" (Anonymous 2020) in many puzzles instead of solution matchstick configuration repeats initial matchstick configuration! In some cases, there is no relationship between the puzzle formulation and its published "solution"! One example is the Puzzle 56. Its formulation (a) initial configuration, (b) published *incorrect* solution, and (c) two omitted *correct* solutions are given in Fig. 2.16.

A mindless error happened in the book "*Matchstick puzzles*" (Di Fred 2015) with the Puzzle 58. Its formulation, initial matchstick formulation, and "solution" are presented in Fig. 2.17.

The error is mindless because of two incredible facts: (1) The published "solution" doesn't contain a single square and is a possible solution to an unknown arithmetic matchstick puzzle with Roman numerals. (2) The Puzzle 58 is a repetition of the Puzzle 55 for which the author did publish a correct solution!

2.5 Possible Cause of This Unfortunate Situation with Recent Books on Matchstick Puzzles

It would be a simplistic approach to think that the only cause is "democratization" of publishing process in which incompetent (and even anonymous) authors have chance to make money by irresponsibly selling errors-filled books on matchstick puzzles using Amazon.

"Democratization" of publishing process is a superficial cause because it is natural consequence of an extended and much deeper cause: cognitive bias which can be called "uncritical thinking", and which commonly leads to erroneous solutions. It was present in the books on matchstick puzzles written by authors with full names and produced by established publishing companies. Here comes an amazing example of "uncritical thinking" of one author and three editors in a scientific book published by Springer (Bailer-Jones 2002, pp. 67–68) is given in Fig. 2.18.

The presence of cognitive bias of "uncritical thinking" among book authors shaped established "matchstick puzzle culture" which is characterized by other three "cognitive biases": "one puzzle—one solution", "formulation change doesn't matter", and "sought squares or triangles are equal-sized". A selection of erroneous solutions due to the cognitive bias of "uncritical thinking", that appeared in books from different epochs, and their possible uses for fostering critical and creative thinking or puzzle solvers, are presented in Chap. 12.

Consider the configuration of matches, shown on figure below, consisting of 12 matches and forming ***three*** squares. The problem to solve is: How can one produce five squares by moving three matches? 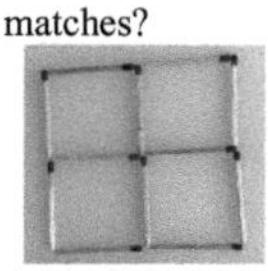	The solution to the puzzle with matches – producing ***five*** squares by moving three matches. 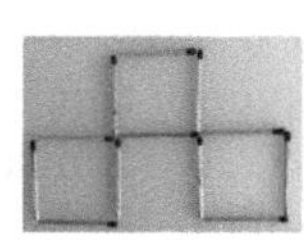

Fig. 2.18 An amazing example of "uncritical thinking" of one author and three editors in a book published by Springer (*Source* Author)

References

Anonymous (2020). *100 matchstick puzzles. Fun family activities.* Ally Creativa.

Arnold, G. (1858). *The Sociable or one thousand and one home amusements.* Dick and Fitzgerald Publisher.

Bailer-Jones, D. M. (2002). Sketches as mental reifications of theoretical scientific treatment. In: Anderson, M., Meyer, B. & Olivier, P. (Editors) *Diagrammatic representation and reasoning.* Springer.

Braun, F. (1876). *Der junge Mathematiker und Naturforscher.* Verlag von Otto Spamer.

Di Fred, A. (2015). *Matchstick puzzles.* CreateSpace Independent Publishing Platform.

Mittenzwey, L. (1880). *Mathematische Kurzweil.* Julius Klinkhardt.

Werner (1865). Die Winterabende unserer Kinder II. Selbstbeschäftigung. *Cornelia. Zeitschrift für häusliche Erziehung, 3*(1), 17–23.

Chapter 3
Matchstick or Toothpick Puzzles in Brain Training and Different Areas of Learning

Abstract This chapter presents a critical analysis of the use of matchstick or toothpick puzzles in brain training and different areas of human learning in search of creativity. Fostering brain health is important in personal life, while fostering creativity is necessary for successful work in many professions. Special attention is given to the use of matchstick or toothpick puzzles in mathematics education. Through documental research, the presence and characteristics puzzles were analyzed in many mathematics textbooks, mainly in teacher's editions. The results of the research, presented in the chapter, show that the use of matchstick or toothpick puzzles is far from adequate. Namely, the number of puzzles used is insufficient and they are prone for didactic improvement. The main failure is the presence of only one solution for multiple-solution puzzles (cognitive bias "one puzzle—one solution").

3.1 Brain Training to Lower Risk of Age-Related Cognitive Decline

Gary Small and Gigi Vorgan recommended that solving toothpick puzzles is one of those useful mental activities that reduce the risk of Alzheimer's disease (Small and Vorgan 2012). Nevertheless, in presenting the solutions to two of toothpick puzzles they, as many other book authors, revealed the cognitive bias "one puzzle—one solution".

The formulation, initial toothpick configuration, and the solution to the first puzzle (Small and Vorgan 2012, p. 132, p. 134) are given in Fig. 3.1.

In fact, this puzzle has five additional, visually different solutions obtained by rotation of the published one. It is very likely, that many readers will find a solution that is different from the published one!

The formulation, initial toothpick configuration, and the verbal description of solution to the second puzzle (Small and Vorgan 2012, p. 132, p. 134) are presented in Fig. 3.2.

Namely, the second puzzle has another solution that is even more creative: Mirror image of the incorrect equation X + I = IX is a correct equation XI = I + X (Fig. 3.3).

J. Slisko, *Fostering Cognitive Mathematics Skills with Matchstick Puzzles*, Mathematics in Mind, https://doi.org/10.1007/978-3-032-14561-1_3

Arrange 12 toothpicks as displayed below, and then try to create three triangles by moving only four of the toothpicks.

Fig. 3.1 The first puzzle of Small and Vorgan that reveals the cognitive bias "one puzzle – one solution" (*Source* Author)

The equation below is incorrect. How can you make it correct without moving a single line? 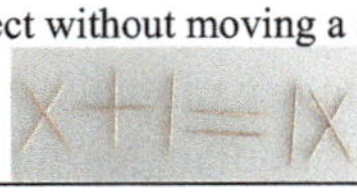	Turn the book upside down – you'll see, 11, 1, and 10 in Roman numerals.

Fig. 3.2 Small and Vorgan presented only one solution to the second puzzle, revealing the cognitive bias "one puzzle—one solution" (*Source* Author)

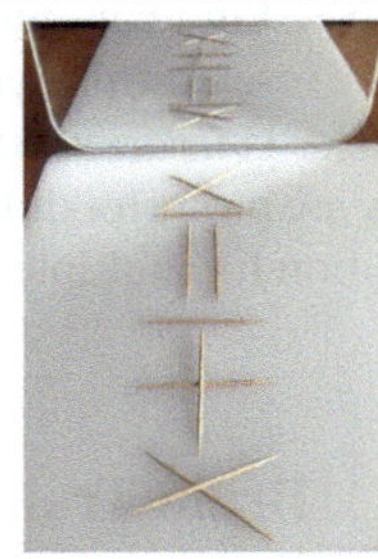

Fig. 3.3 Additional solution to the second puzzle of Small and Vorgan (*Source* Author)

Cognitive bias "one puzzle—one solution" is also present in a recent book by Small and Vorgan in which they write (Small and Vorgan 2023, p. 6):

> Accumulating scientific evidence supports the idea that brain workouts – *doing puzzles*, taking classes, playing games, reading books, or even engaging in stimulating conversations – can lower a person's risk for age-related cognitive decline.

One example of that cognitive bias is present in an arithmetic puzzle with Arabic numerals. Its formulation **(a)** initial toothpick configuration, **(b)** given solution, and **(c)** omitted additional solution are shown in Fig. 3.4.

A more serious cognitive bias of "uncritical thinking" happened in a geometric toothpick puzzle (Small and Vorgan 2023, p. 22, p. 139). Its formulation, initial toothpick configuration, and the "solution" are given in Fig. 3.5.

The solution is erroneous because, beside the two sought rectangles ("1 × 2" and "1 × 3"), the remaining toothpicks display additionally one "1 × 1" square and one "2 × 3" rectangle.

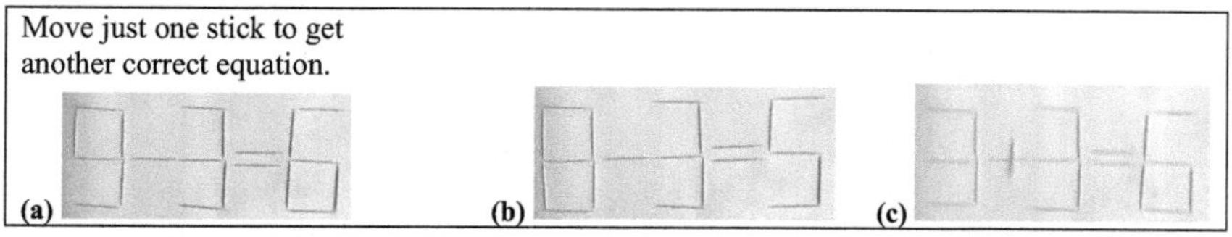

Fig. 3.4 Arithmetic toothpick puzzle in which Small and Vorgan reveled cognitive bias "one puzzle—one solution" (*Source* Author)

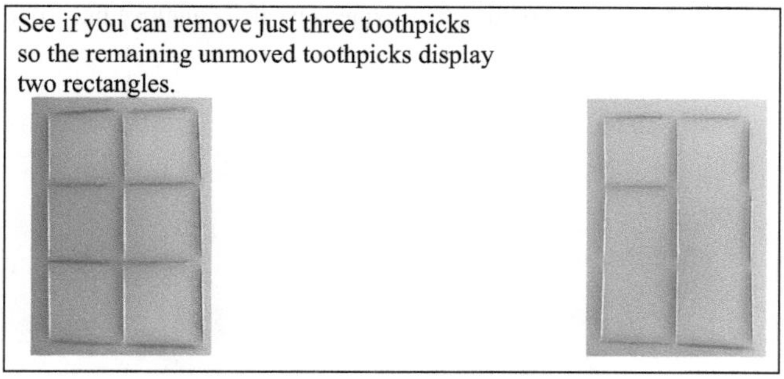

Fig. 3.5 Arithmetic toothpick puzzle in which Small and Vorgan reveled cognitive bias "uncritical thinking" (*Source* Author)

3.2 Fostering Creativity with Solving Matchstick Puzzles

Geometric matchstick puzzles with multiple solutions are ideal tasks for fostering creative thinking of puzzle solvers. It is a big surprise to find out that such potential was very rarely used in books who were supposedly written to promote creativity. Even authors who presented two different classes of solutions to a matchstick puzzle for each class of solution gave only one of possible solutions (Evans 1990, p. 23, p. 37).

In the book with highly attractive title "*Five-star mind. Games and exercises to stimulate your creativity and imagination*" (Wujec 1995, p. 127), five matchstick puzzles are included with the aim "to stimulate … creativity and imagination". All five puzzles were formulated for the same initial, unshown matchstick configuration (a grid three by three made from 24 matchsticks). The first and fourth puzzles have only one solution. The second, third, and fifth puzzles are well-known multiple-solution puzzle.

Nevertheless, the author's solutions to the third and fifth puzzles show the presence of cognitive bias "one puzzle—one solution". That cognitive bias will be illustrated in the case of the third puzzle. Its formulation, referential matchstick configuration (not presented visually in the book!), and single solution (Wujec 1995, p. 242) are given in Fig. 3.6.

The third puzzle has various additional fundamentally different solutions which weren't known to the author. Three of those solutions are shown in Fig. 3.7.

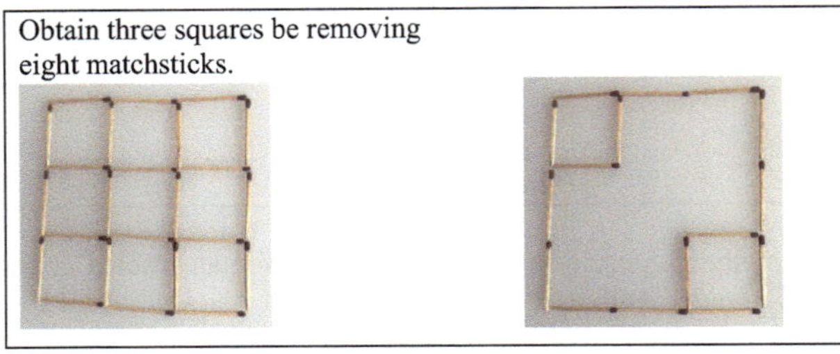

Fig. 3.6 Geometric matchstick puzzle in which Wujec revealed cognitive bias "one puzzle—one solution" (*Source* Author)

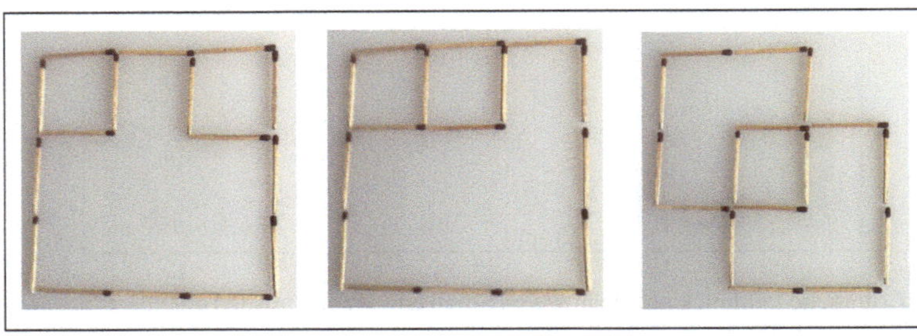

Fig. 3.7 Three additional fundamentally different solutions to third puzzle unknown to Wujec (*Source* Author)

The solution to the second puzzle (Wujec 1995, p. 127, p. 242) shows an amazing case of cognitive bias "uncritical thinking". The formulation of the second puzzle, its referential matchstick configuration, and single "solution" are presented in Fig. 3.8.

"Uncritical thinking" is revealed by the fact that the author didn't notice that the "solution" presented to the readers contains ***six identical squares*** instead of ***sought five identical squares***.

Three of various correct solutions to the second puzzle with five identical squares are shown in Fig. 3.9.

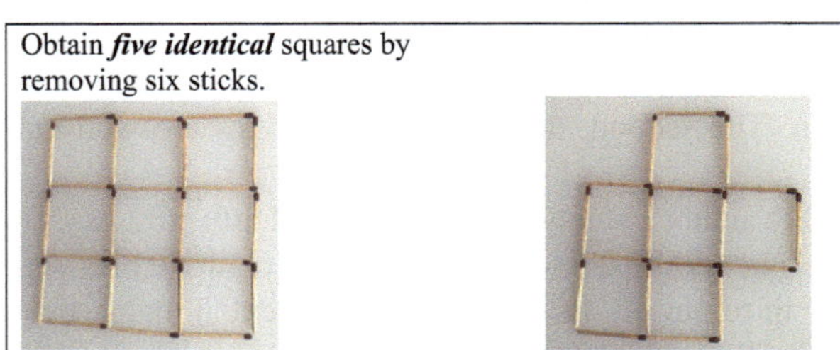

Fig. 3.8 The second geometric matchstick puzzle with "solution" in which Wujec presented cognitive bias "uncritical thinking" (*Source* Author)

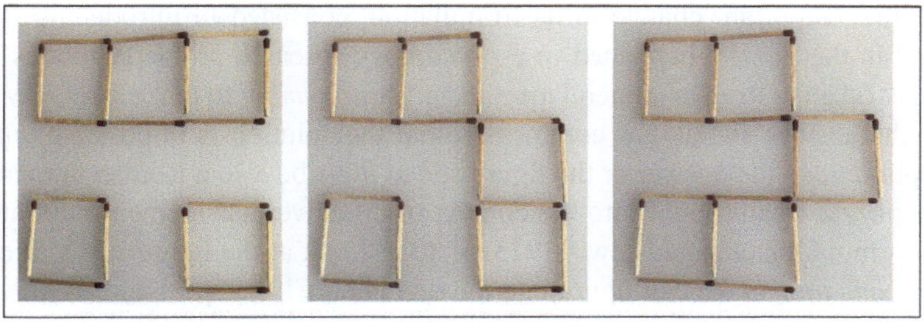

Fig. 3.9 Three additional fundamentally different correct solutions to second puzzle unknown to Wujec (*Source* Author)

Other book authors, whose aim is to illustrate and promote creative thinking and mental fitness in the readers, also don't escape cognitive biases of "one puzzle—one solution" and "uncritical thinking".

Baker presented only a single solution to three multiple-solution geometric toothpick puzzles (Baker 1962, p. 15, p. 230; p. 89, p. 248; p. 91, p. 251), Carter and Russel did the same in a three-solution geometric matchstick puzzle (Carter and Russel 2002, p. 45, p. 183), and an anonymous author presented only one solution to a four-solution geometric puzzle (Anonymous 1994, puzzle 2, p. 9, p. 134).

It is very unfortunate that there are books written to develop creativity which reveal "uncritical thinking", presenting matchstick puzzles with erroneous solution (Glover 1980, p. 163).

3.3 Usefulness of Puzzles for Success in Various Professional Works

Matchstick puzzles, along with verbal puzzles, are used in various books whose authors seek to inform readers or course takers (with illustrative examples) about some personal abilities (out-of-box thinking, visual intelligence, problem-solving), which are necessary for successful professional work in business communication, management, and computing and visualization design.

It is interesting to show how presentation form of a very old and popular verbal puzzle and its solution were changing while it was used in different contexts (entertainment, research, business communication, innovation thinking, management skills).

The puzzle "What must you add to nine to make six?" was published in 1851 (Anonymous 1851, p. 72) with the answer "The letter S; for as IX is nine, SIX make six" (Anonymous 1851, p. 92). It was repeated later with the same answer in many posterior books (Anonymous 1861, p. 38; Lemon 1890, p. 91; Loyd 1914, p. 104; Ordway 1915, p. 62; Hindman 1963, p. 143).

The first important change in its formulation, consisted in introducing Roman numeral in the puzzle, happened in a psychology research article (Gibbons 1965, p. 205): "Take the Roman numeral nine and *add one symbol* to make it six". Without giving the answer, the author commented that the solution is impossible with the assumption that "the answer or the figure added must be also a Roman numeral".

The puzzle was brought into the context of effective management with another subtle formulation change (Lewis 1975, p. 80): "Look at the figure … IX. Now, by *drawing one line*, make that figure into the number six". The problem "is solved by *drawing the letter S* in front of the IX". In the context of business communication, other authors repeated the puzzle and the solution (Huseman et al. 1981, p. 35; Munter 1987, pp. 11–12).

As an example of innovative thinking, von Oech published the second solution of the puzzle (Von Oech 1983, p. 9):

> Another solution might be to add the line "6" after IX. Then you get IX6, or one times six. Here the "X" no longer represents "10" or the English letter "X" but rather the multiplication sign.

The puzzle was brought into the context of management skills development to illustrate how problems are solved creatively (Whetten and Cameron 1995, p. 193; Whetten et al. 1996, p. 34). Its formulation was: "Below is the Roman numeral 9. By adding only a single line, turn it into a 6".

The readers were given three solutions (Whetten and Cameron 1995, p. 693; Whetten et al. 1996, p. 94): "S"IX or "6" or XI.

While the first solution is the known one, the second is an improper presentation of von Oech's solution IX6. The third solution shows the cognitive bias of "uncritical thinking". Namely, while "S" and "6" can be considered as "as a single line", "XI" isn't prone to a such consideration. In addition, as in the second solution, the readers aren't told where the "line" XI should be placed (before or after IX).

An original matchstick puzzle was used to show the importance of creative intuition for management leadership (Daft 2008, p. 490). Its formulation, matchstick configuration, and verbal solution are given in Fig. 3.10.

In the context of managers preparation, a known matchstick puzzle was used as an example of simplifying a diagram or taking some complexity out of it (Plenert 1995, pp. 321–322, p. 329). Its formulation, initial matchstick configuration, and the solution are in Fig. 3.11.

In the illustration below, remove three matches to leave four. 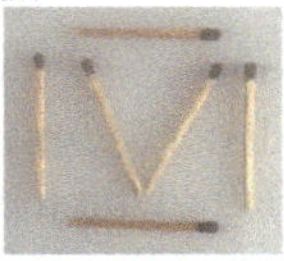	The answer… depends on how you interpret the word "four". Rather than counting four matches, remove the matches at the top, bottom, and right and the answer is obvious – the Roman numeral IV.

Fig. 3.10 Matchstick puzzle for showing the importance of creative intuition (*Source* Author)

Take sixteen matchsticks and arrange them, as shown below, in five equal-size squares. By moving only two of the matches, rearrange the diagram so that all sixteen matches are used but we are left with only four equal-size squares.

Fig. 3.11 Matchstick puzzle for showing the simplification of a diagram (*Source* Author)

The same matchstick puzzle and single solution with equal-sized squares appeared also in the context computing and visualization design (Roberts et al. 2017, pp. 70–71). Nevertheless, as restriction of squares equality was omitted, the authors shown cognitive bias of "formulation change doesn't matter" and didn't present any of possible solutions with unequal-sized squares.

3.4 Matchstick or Toothpick Puzzles in Mathematics Education

In actual mathematics education, textbook authors use matchstick (or toothpick or straw or stick or crayon) puzzles in different forms. Less frequent use of the word "matchstick" is related to the understandable authors' fear that pupils, if asked to "do mathematics" with matchsticks, might cause fire in school.

Different forms of usage of matchstick or toothpick puzzles in mathematics textbooks are in many cases inadequate, missing to give pupils and students the opportunities to practice cognitive skills of creative visual thinking and combinatorial reasoning in finding different solutions. That situation is surprising because mathematics educators in the role of textbook authors, some with PhD in mathematics education, should be better in avoiding many cognitive biases of those book authors on matchstick puzzle without formal mathematics background at the university level.

3.4.1 Puzzles Without Given Solutions

The first matchstick puzzle, described in Chap. 2, entered the world of mathematics education when it was only 10 years old! Edward Brooks, recognized pedagogue and professor of mathematics at the *Pennsylvania State Normal School,* included it in his book "*Methods of teaching mental arithmetic*" (Brooks 1860). The puzzle was part of the last section "*Social arithmetic*", which contained 43 problems and puzzles whose aim was to provide possibilities for funny social activities near the fireplace

Place 17 little sticks – matches, for instance – making 6 equal squares, as in the margin. Then remove 5 sticks, and leave 3 perfect squares of the same size.

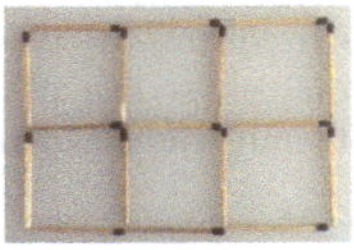

Fig. 3.12 The first matchstick puzzle used in mathematics education by Brooks (*Source* Author)

Fig. 3.13 A multi-solution stick puzzle without given solution (*Source* Author)

Move 3 sticks to make 5 squares that are the same size.

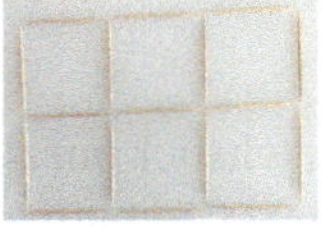

in winter nights. Brooks' version of the first matchstick puzzle (Brooks 1860, p. 74) is given in Fig. 3.12.

Although in some other problems and puzzles Brooks provided the answers, due to an unknown reason, for the single matchstick puzzle the answer was omitted.

It is strange that even today some textbook authors repeat the same approach when using a single, very good multiple-solution puzzle (Duncan et al. 1983, p. 75), presented in Fig. 3.13.

The same happened with a five-part toothpick puzzle in a book (Long and Leland 2002, p. 44) and for a four-part stick puzzle even in a journal article (Anonymous 1959, p. 179).

3.4.2 *Puzzles with Single Solutions*

The next inadequate use of puzzles in mathematics education is to present only those puzzles with single solutions. By doing so, authors of mathematics textbook foster a known students' false belief "all problems have only one solution". Annebelle R. Bucknam started this approach in a pedagogical journal for primary teachers, recommending toothpick puzzles as "especially good to use in rural schools during stormy winter noons after lunch is over" (Bucknam 1917, p. 46). The chosen geometric toothpick puzzle and its solution are given in Fig. 3.14.

Place fifteen toothpicks to form five squares. The puzzle is to remove three toothpicks, and leave three perfect squares. Let the children try this until they solve it.

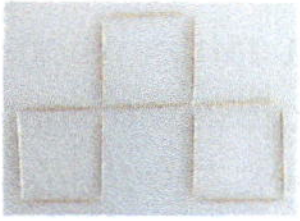

Fig. 3.14 A geometric toothpick puzzle with single solution (*Source* Author)

The figure at the right shows twelve toothpicks arranged to form three squares. How can you form five squares by moving only three toothpicks?

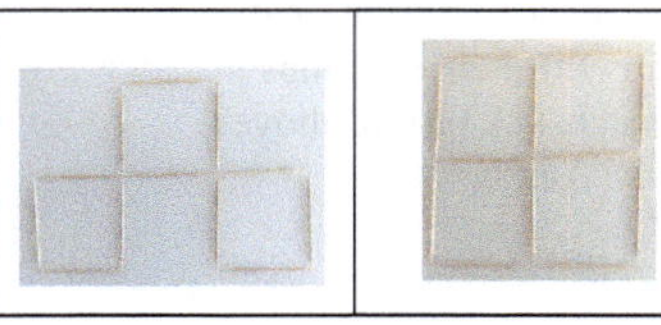

Fig. 3.15 The first example of geometric toothpick puzzle with single solution (*Source* Author)

In the diagram 11 toothpicks make three squares. Move three toothpicks to produce only two squares.

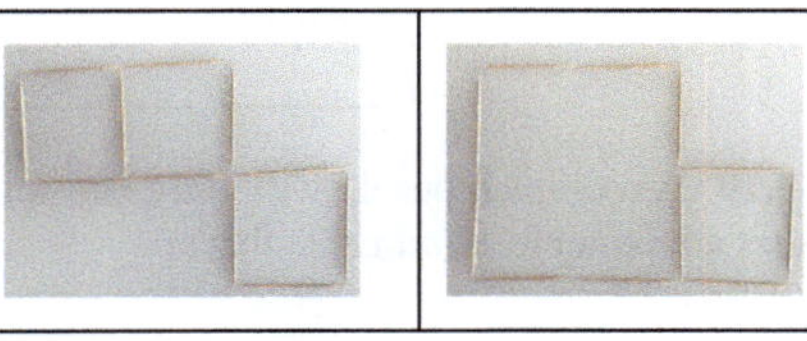

Fig. 3.16 The second example of geometric toothpick puzzle with single solution (*Source* Author)

Two examples of such a use of toothpick puzzles come follow. The first example (Lenchner 2005, p. 26) is in Fig. 3.15.

The second example (Mickelson and Mickelson 2020, p. 341, p. 388) is in Fig. 3.16.

3.4.3 *Puzzles with Multiple Solutions*

Many geometric matchstick or toothpick puzzles have multiple solutions, which can be divided into two groups:

(1) visually different solutions between which rotational or reflection relationships exist;
(2) fundamentally different solution between which rotational or reflection relationships don't exist.

There are two approaches of textbook authors to visually different solutions to puzzles. The most common, and the worst pedagogical approach, is to publish *only one* of possible visually different solutions (cognitive bias "one puzzle—one solution"). Formulation, initial matchstick configuration, and solution of a puzzle in such an approach (Thyer 1993, p. 77, p. 165) are given in Fig. 3.17.

This puzzle has three additional, visually different solutions (Fig. 3.18).

Fig. 3.17 A matchstick puzzle with only one solution published (*Source* Author)

Remove 2 matchsticks so that only 2 squares remain.

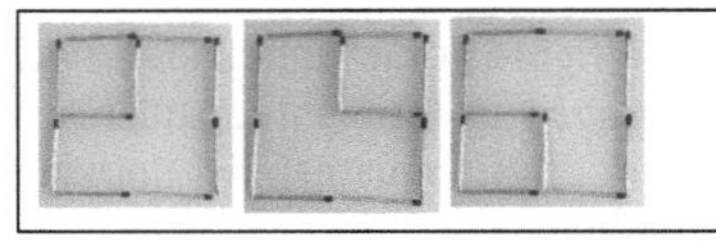

Fig. 3.18 Three additional, visually different solutions to the matchstick puzzle above (*Source* Author)

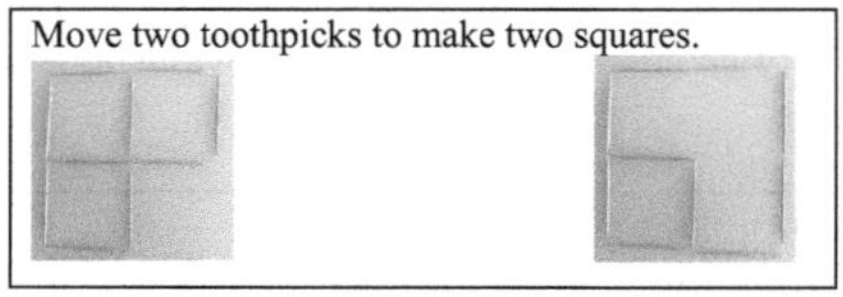

Fig. 3.19 A toothpick puzzle with only one solution published but with the warning that "other answers are possible" (*Source* Author)

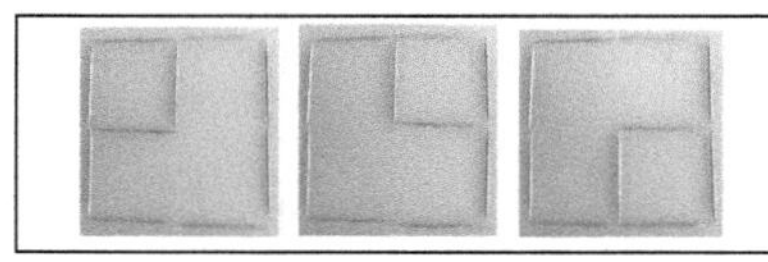

Fig. 3.20 Three additional, visually different solutions to the matchstick puzzle above (*Source* Author)

A slightly better, but still pedagogically insufficient approach is to publish one solution, warning teachers that "other answers are possible", without specifying their number.

Formulation, initial toothpick configuration, and solution to a puzzle in that better approach (Bornhold et al. 1982, p. 177) are given in Fig. 3.19.

"Other possible answers" are, of course, three visually different solutions (Fig. 3.20).

Pedagogically adequate approach for fostering visual creativity would be to inform students (and teachers!) about the exact number of visually different solutions to matchstick or toothpick puzzles, giving them two tasks: (1) find all visually different solutions, and (2) establish rotation and reflection relationships between these different solutions.

Mathematics textbook authors also reveal cognitive bias "one puzzle—one solution" in the case of matchstick or toothpick puzzles with fundamentally different solutions in which authors published only one of them. Here come two illustrative examples of many such multiple-solution puzzles.

The formulation, (a) initial toothpick configuration, (b) given single solution, and (c) two omitted additional, fundamentally different solutions to the first puzzle (Anonymous 1998, p. 67) are given in Fig. 3.21.

The formulation, initial matchstick configuration, and given single solution to the second puzzle (Joshua 1994, p. 6, p. 38) are presented in Fig. 3.22.

The second puzzle has additional 14 fundamentally different solutions, one with four squares of equal size and 13 with four unequal squares (one 2×2 square and three 1×1 squares). These solutions are given in Fig. 3.23.

Your classmate puts toothpicks together to form 5 squares. He asks you to remove 2 toothpicks to produce 4 squares. Which toothpicks should you remove?

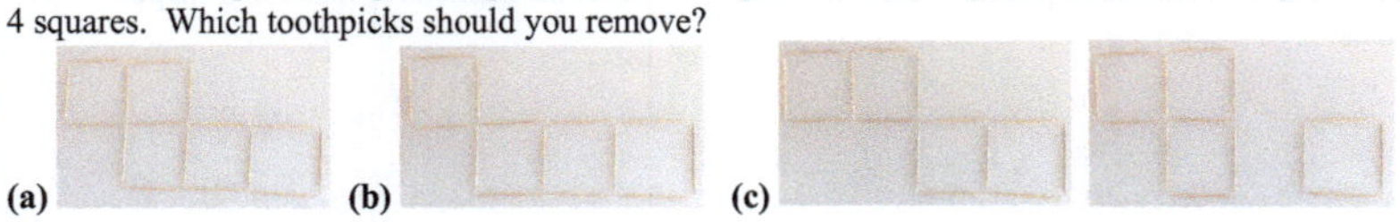

Fig. 3.21 Formulation, initial toothpick configuration, given solution, and two omitted visually different solutions to the first puzzle (*Source* Author)

Remove 4 matches to leave 4 squares.

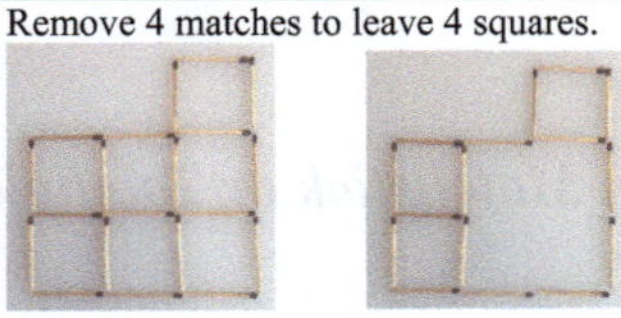

Fig. 3.22 Formulation, initial matchstick configuration, and given solution to the second puzzle (*Source* Author)

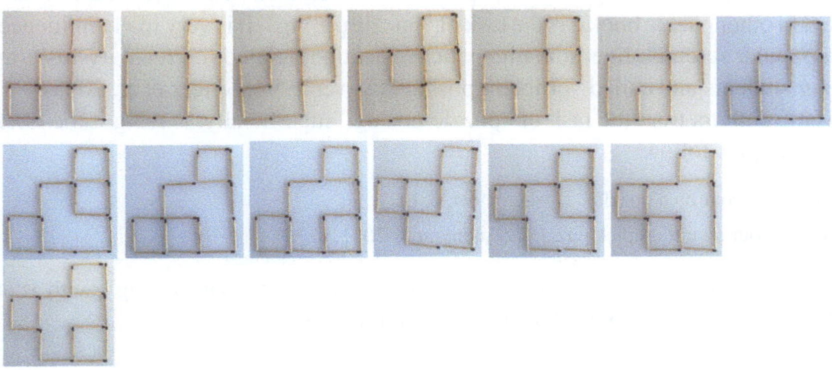

Fig. 3.23 Fourteen omitted fundamentally different solutions to the second puzzle (*Source* Author)

It is important to mention that the same author for the previous puzzle ("Remove 2 matches to leave 5 equal squares") published 4 solutions, three of them fundamentally different (Joshua 1994, p. 6, p. 38)! This fact shows that the level of creative puzzle solving isn't constant and can change from puzzle to puzzle.

Additional examples of the same cognitive bias "one puzzle—one solution" can be found in other mathematics textbooks (Hoffer et al. 1992, Puzzle 4, p. 47; Lynch and Olmstead 1998, Puzzles 8 and 9, p. 22; Overholt and Kincheloe 2010, Puzzle 23, pp. 370–371).

Create some "taking" matchstick puzzles based on these arrangements:  or 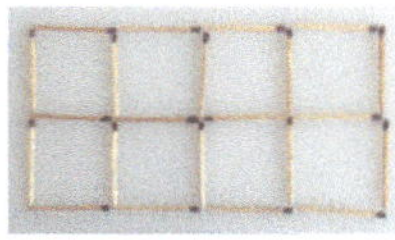 (Kirby, 1994, p. 16)	Use matchsticks to make this shape. 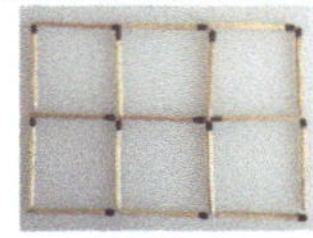 Can you take away 5 matches and leave only three squares (all the same size)? Try to make up other matchstick puzzles. (Banwell at al., 1975, p. 109)

Fig. 3.24 Two examples of pedagogical treatments of "puzzle posing" with scaffolding approach (*Source* Author)

3.4.4 Make Your Own Matchstick or Toothpick Puzzle

Inadequate treatments of matchstick or toothpick puzzles in mathematics textbooks are somehow an uncritical acceptance of treatments of matchstick puzzles in puzzle books not written for strict educational use. Pedagogical inadequacy of the later books will be additionally documented in posterior chapters.

The only good novelty, which appears in mathematics textbooks and is totally absent in other puzzle books, is the task of "puzzle posing". It is very likely influenced by a popular research line in mathematics education, related to creativity fostering among students and called "problem posing" (Silver 1994; Singer et al. 2015).

Some authors treat the task of "puzzle posing" superficially, leaving it too open. Two examples follow:

> "Make up your own puzzle." (Joshua 1994, p. 6).
>
> "Make your own toothpick puzzle." (Anonymous 2001, p. 18).

Two examples of better pedagogical treatments of "puzzle posing" with scaffolding approach (Anghileri 2006) are given in Fig. 3.24.

References

Anghileri, J. (2006). Scaffolding practices that enhance mathematics learning. *Journal of Mathematics Teacher Education*, *9*(1), 33–52.

Anonymous (1851). *The book of riddles*. Darton and Company.

Anonymous (1861). *Do you give it up? A collection of the most amusing conundrums, riddles, etc. of the day*. Lockwood and Company.

Anonymous (1959). Stick puzzle. *Arithmetic Teacher, 6*(4), 179.

Anonymous (1994). *Mindpower. Develop your creative skills*. Time-Life Books.

Anonymous (1998). *Houghton Mifflin Math Central: Enrichment*. Teacher's annotated edition. Houghton Mifflin.

Anonymous (2001). *Enrichment masters. Glencoe geometry. Integration, application, connection.* Glencoe/McGraw-Hill.

Baker, S. S. (1962). *Your key to creative thinking. How to get more and better ideas*. Harper & Row.

Banwell, C. S., Hiscocks, J. E., Paling, D., Saunders, K. D., Wardle, M. E. & Weeks, C. J. (1975). *Oxford comprehensive mathematics. Book 1. A secondary course for mixed abilities.* Oxford University Press.

Bothe, D. R. (2002). *Reducing process variation.* Landmark Publishing Company.

Bornhold, D. L., Gutcher, R., Tossel, S. & Traynor, C. (1982). *Starting Points in Mathematics.* Level 3. Teacher's edition. Ginn and Company.

Brooks, E. (1860). *Methods of teaching mental arithmetic, and Key to the normal mental arithmetic, containing also many suggestions and methods for arithmetical contractions, and a collection of problems of an interesting and amusing character, for class exercise.* Sower, Barnes & Company.

Bucknam, A. R. (1917). Puzzles for play periods. *Primary Education. A Monthly Journal for Primary Teachers*, XXV(1), 46.

Carter, P. & Russel, K (2002). *Maximize your brain power. 1000 new ways to boost your mental fitness.* John Wiley and Sons.

Daft, R. L. (2008). *Leadership for Managers.* Cengage Learning.

Duncan, E. R., Quasta, W. G., Cole, W. L., Sparks, T. M. & Haubner, M. A. (1983). *Houghton Mifflin Mathematics.* Houghton Mifflin.

Evans, P. (1990). *The keys to creativity.* Grafton Books.

Gibbons, D. (1965). Set-breaking as a learned response. *Psychological Reports, 17*, 203–208.

Glover, J. A. (1980). *Becoming a more creative person.* Prentice-Hall.

Hindman, D. A. (1963). *1800 riddles, enigmas and conundrums.* Dover Publications.

Hoffer, A. R., Lodholz, R. D., Johnson, M. L., Musser, G. L., Leinwand, S. J. (1992). Mathematics in action. Teacher's edition. Macmillan/McGraw-Hill.

Huseman, R. C., Lahiff, J. M. & Hatfield, J. D. (1981). *Business communication. Strategies and skills.* The Dryden Press.

Joshua, A. (1994). *Maths Challenges.* Graded problems for 10–13-year-old. Simon & Shuster Education.

Kirkby, D. (1994). *Squares and triangles.* Folens.

Lemon, D. (1890). *Everybody's illustrated book of puzles.* Saxon and Company,

Lenchner, G. (2005). *Creative problem solving in school mathematics. A handbook for teachers, parents, students, and other interested people.* Second edition. Mathematical Olympiads for Elementary and Middle Schools.

Lewis, P. V. (1975). *Organizational communication. The essence of effective management.* Grid Incorporation.

Long, F. & Leland, G. (2002). *Summer bridge middle school.* 6th to 7th grade. Rainbow Bridge Publication.

Loyd, S. (1914). *Sam Loyd's cyclopedia of 5000 puzzles, tricks, and conundrums with answers.* The Lamb Publishing Company.

Lynch, C. & Olmstead, E. (1998). *South-Western math matters. An integrated approach.* Book 3. Teacher's annotated edition. South-Western Educational Publishing.

Mickelson, R. J. & Mickelson, P. (2020). *Theory and Problems for Mathematics 8.* Third Edition. Crescent Beach Publishing.

Munter, M. (1987). *Business communication. Strategy and skill.* Prentice-Hall.

Ordway, E. B. (1915). *The handbook of conundrums.* George Sully and Company.

Overholt, J. & Kincheloe, L. (2010). *Math Wise! Over 100 hands-on activities that promote real math understanding. Grades K-8.* Second Edition. Jossey-Bass.

Plenert, G. J. (1995). *World class manager. Olympic quality performance in the new global economy.* Prima Publishing.

Roberts, J. C., Headleand, C. J. &. & Ritsos, P. D. (2017). *Five design-sheets: Creative design and sketching for computing and visualization.* Springer.

Small, G. & Vorgan, G. (2012). The Alzheimer's prevention program. Keep your brain healthy for the rest of your life. Workman Publishing.

Small, G. & Vorgan, G. (2023). *Brain games to exercise your mind: Protect your brain from memory loss and other age-related disorders: 90 puzzles, logic riddles & brain teasers to exercise your mind*. Humanix Books.

Silver, E. A. (1994). On mathematical problem posing. *For the learning of mathematics, 14*(1), 19–28.

Singer, F. M., Ellerton, N. F. & Cai, J. (2015). *Mathematical problem posing*. Springer.

Thyer, D. (1993). *Mathematical enrichment exercises*. A teacher's guide. Cassel.

Von Oech, R. (1983). *A whack on the side of the head. How to unlock your mind for innovation.* Warner Books.

Whetten, D. A. & Cameron, K. (1995). *Developing management skills*. Third edition. HarperCollins Publishers.

Whetten, D. A., Cameron, K. & Woods, M. (1996). *Effective problem-solving*. HarperCollins Publishers.

Wujec, T. (1995). *Five-star mind. Games and exercises to stimulate your creativity and imagination.* Doubleday.

Chapter 4
Different Investigations Related to Solving Matchstick Puzzles

Abstract This chapter presents the results of a few investigations carried out with uses of matchstick puzzles. George Katona (1901–1981) explored experimentally how two different pedagogies influence the performances of matchstick solvers involved. The first pedagogy consists in showing the one solution to the puzzle, fostering memorization of the solution. The second pedagogy is to help solvers grasp the principle that leads to the solution, fostering solving process understanding and skills. Katona's results show that the second pedagogy is better in solving new matchstick puzzles. In addition, these results contradict beliefs that there aren't general rules for solving matchstick puzzles or that they can be solved only by a trial-and-error approach. Joy Paul Guilford (1897–1987) used various types of matchstick puzzles to explore divergent thinking as an important part of creativity. He also mentioned cognitive bias "sought squares must be equal" that makes solving some puzzles impossible. Various investigations demonstrated existence of a theoretical base for predicting difficulty levels of different arithmetic matchstick puzzles with Roman numerals. These predictions were verified experimentally observing the number of correct solutions and time needed to find them.

4.1 First Investigations on Solving Matchstick Puzzles

Psychological research on how persons solve puzzles was started by John Hudson Ballard (Ballard 1915). Among the puzzles used in that research was also the first matchstick puzzle, presented in Chap. 2. Ballard's formulation, initial matchsticks configuration, and solution (Ballard 1915, p. 30, p. 95) are given in Fig. 4.1.

From eleven participants with very diverse ages, in one part of the research, only five were able to find the solution. It is important to mention that Ballard said that the given solution was the only possible one, not noting the existence of its rotation or reflection, visually different solution variant.

Mittenzwey's popular matchstick puzzle "*Form 4 congruent triangles from 6 matchsticks*" (Chap. 2, Fig. 2.15) was used in different experimental psychological investigations. Researchers were interested in knowing more about existence and

J. Slisko, *Fostering Cognitive Mathematics Skills with Matchstick Puzzles*, Mathematics in Mind, https://doi.org/10.1007/978-3-032-14561-1_4

Seventeen matches are arranged in six contiguous squares... The problem is to remove only five matches and leave intact three of the original squares.

Fig. 4.1 The matchstick puzzle and its solution used in Ballard's research (*Source* Author).

nature of "insight" moments (also known as Eureka or AHA moments) in which persons, after a few unsuccessful tries, suddenly get a new idea how a problem can be solved. Two of these investigations will be briefly presented and commented.

In her research, Mary Elizabeth Bulbrook gave to unspecified number of participants the following instruction (Bulbrook, 1932, p. 430):

> You have presented to you six matches. Construct of them four equilateral triangles without breaking the matches. Comment aloud.

Unsuccessful tries of all participants were caused by the assumption that all triangles must lie in a plane. Only three participants, observing chance position of a matchstick touching the plane by one of its two ends, got new idea how to find the solution in three dimensions. All other participants (except one) concluded it impossible to construct the required triangles.

John W. Reid was interested in knowing whether six verbal aids might help in the process of reaching insight moment after unsuccessful 5-minute try (Reid 1951). He gave to 35 participants the following instruction (Reid 1951, p. 55):

> Make four equilateral triangles all match length size with these six matches. There are no restrictions other than that the triangles must all be match length size. Any other restrictions are made by you, so don't make any!

Ten participants were able to solve the problem during five minutes. First two verbal aids weren't useful to any of 25 left participants. Third verbal aid helped eight participants to find the solution. After heard the fourth verbal aid, the solution was found by additional four participants. Fifth and sixth verbal aids were useful to one and five participants.

Seven participants were unable to find the solution. They had a "fixation approach" and "worked with energy, yet producing variation and repetition of countless plane figures".

Formation of four equilateral triangles from six matchsticks is far from a trivial matchstick manipulation (Fig. 4.2).

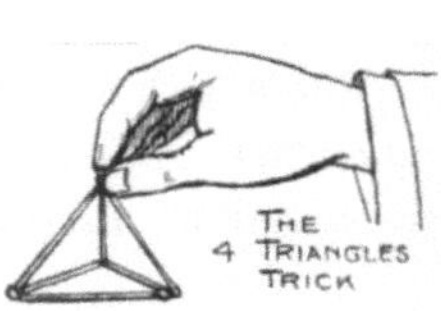

Fig. 4.2 The tridimensional solution of the four-triangle matchstick puzzle (*Source* Anonymous 1920).

Being so, it is surprising that neither Bulbrook nor Reid considered necessary to describe the forms in which successful participants practically presented their solutions!

4.2 Katona's Investigation on Solving Matchstick Puzzles

George Katona (1901–1981) was an American psychologist of Hungarian origin who wrote an important research-based book about psychology of learning and teaching problem-solving by using matchstick puzzles (Katona 1940). Until today, nobody carried out such a big number of related experiments. Katona's main research goal was to find the best teaching method for successful delayed transfer. It means that the participants had to show the following: what was learned in teaching phase leads to a good performance in later testing phase of solving known and new puzzles.

4.2.1 Two Teaching Methods for Solving Matchstick Puzzles

Katona explored the learning effects of two different teaching methods. In his first experiments, the participants in both teaching methods were given two matchstick tasks (Fig. 4.3).

Both tasks had the same instruction: Make four squares out of the five squares by changing the positions of three matchsticks.

It is important to mention that Katona knew that the Task 1 had three other and Task 9 two other correct solutions and that some of these solutions involved the separation of the original figure into two noncontiguous parts. The same knowledge can be noted in other puzzles used in investigation. Katona obviously escaped common cognitive bias "one puzzle—one solution" which is revealed by many book and mathematics textbook authors.

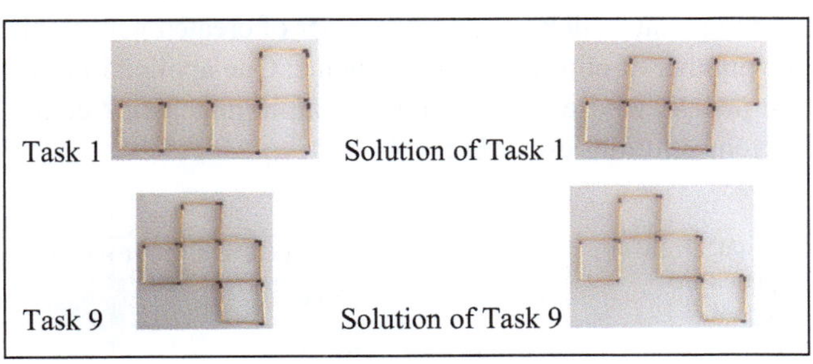

Fig. 4.3 Two matchstick tasks with given one solution used initially by Katona in two different teaching methods (*Source* Author).

Teaching Method I (TM I) in which *the solution is taught*. It had three steps. First step: Task 1 is given to the participants, and it is not solved within two minutes. Second step: Task 9 is given, and it is not solved within two minutes. Third step: The solution of Task 1 is shown to the participants. Test: Task 9 is given again for three minutes.

The learning results of this teaching method were negative. Nobody was able to solve Task 9 after seeing the solution of Task 1. Additionally, a week later four participants were retested, and only one was able to solve Taks 1.

Facing these negative results, and mixed results of various experiments in which participants were solving Tasks 1 and 9 without assistance, Katona decided to design a second teaching method based on the explanation of the mathematical principle which forms the basis of the matchstick tasks.

Teaching Method II (TM II) in which *the principle is taught*. The first and the second steps are the same as in Method I. The third step is different. Task remains before the participants while the following explanation is given:

> Here are five squares composed of sixteen matchsticks. We want to change these five squares into four similar squares. Since we have sixteen matchsticks and want four squares, each square must have its four independent matchsticks, which should be sides of any other square at the same time. Therefore, all matchsticks with a double function, that is, limiting two squares at the same time, must be changed into matchsticks with a single function (limiting one square only).

The basic idea of this "arithmetical principle" is essentially equal to the idea of Braun's strategy for solving geometric matchstick puzzles which was presented in Chap. 2.

For a few participants, this teaching method was good. Namely, they were able solve Task 1 and Task 9 and did the same after 8 days. Nevertheless, for many participants the method wasn't good because they didn't succeed in solving Task 1 or Task 9. For some of them Method II was even less effective than Method I!

Considering the wording of "arithmetical principle" too abstract, Katona designed concrete "structural principle". By it, puzzle solvers are taught to count structural elements (squares or triangles) in initial and final matchstick configurations to find sought solution. If the matchstick puzzle asks to increase the number of structural elements by moving matchsticks, then the number of created new elements must be bigger than the number of eliminated old elements. One application example of this principle is the puzzle, whose formulation, (a) initial matchstick configuration and (b) one of different solutions are in Fig. 4.4.

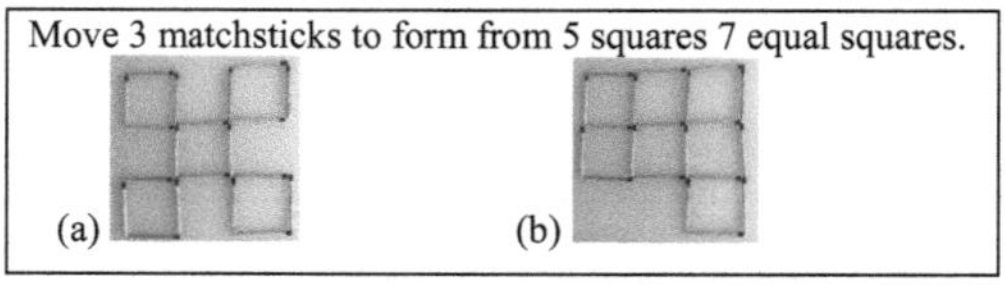

Fig. 4.4 Katona's puzzle in which 5 equal squares are converted into 7 equal squares by moving 3 matchsticks (*Source* Author).

As the number of squares must increase by two, three matchsticks should be moved in such a way that one old square is eliminated, and three new squares are created.

Inversely, if the matchstick puzzle asks to reduce the number of structural elements by moving matchsticks, then the number of created new elements must be smaller than the number of eliminated old elements. One application example of this principle is the puzzle, whose formulation, (a) initial matchstick configuration and (b) one of different solutions are in Fig. 4.5.

As the number of squares must decrease by one, three matchsticks should be moved in such a way that two old squares are eliminated, and one new square is created.

Application of "structural principle" in TM II gave better results in participants' learning than TM I. In one comparative experiment, 25 participants taught by TM II were able to find 12 correct solutions of a new puzzle, while 27 participants taught by TM I found only 5 correct solutions.

Katona's "structural principle" will be called in posterior chapters "Katona's strategy" for solving matchstick puzzles.

Katona's investigations demonstrated that there are didactic possibilities to teach students strategic thinking in solving matchstick puzzles. His results show that are false those affirmations in mathematics education which claim that there isn't a general rule solving matchstick or toothpick puzzles or that these puzzles can be solved only by trial and error:

> It is impossible to state any general rules for solving puzzles with matches but the fun and the challenge remain. (Vollmar et al. 2008, p. 84)
>
> This technique helps you solve a puzzle when you don't even know where to start! When you aren't sure how to begin solving a problem, use *trial and error*.
>
> This problem-solving method involves trying something - anything - even if you don't think it will work, and then looking carefully at the result to see what happened. If it isn't what you wanted, go back to beginning, and try something different. Keep trying things (and keep carefully looking at the result!) until you either find the solution or begin to get an idea of how problem works, which may lead to a more efficient way to solve it (Rapport and Yoder 2017, p. 93)

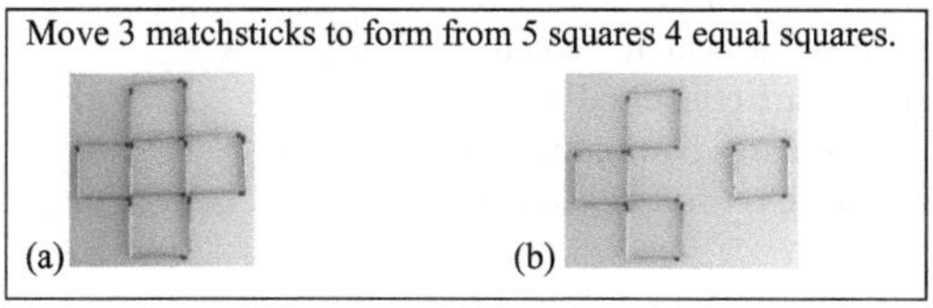

Fig. 4.5 Katona's puzzle in which 5 equal squares are converted into 4 equal squares by moving 3 matchsticks (*Source* Author).

4.3 Guilford's Tests of Creativity by Using Matchstick Puzzles

Joy Paul Guilford (1897–1987) was an American psychologist widely known for his extensive theoretical and experimental research on the nature of human intelligence (Guilford 1959, 1967).

Guilford was also interested in divergent thinking as an important part of creativity (Guilford. 1967). To evaluate this aspect of intelligence, he designed paper-and-pencil tests with different types of "match problems". These tests contained in one part known matchstick puzzles with squares and triangles. In these puzzles, sought final configurations are obtained, not by *moving* matchsticks as in Katona's puzzles, but by *removing* required number of matchsticks. Here come two popular examples of Guilford's "match problems" from the first part (Guilford 1967, pp. 152–153).

Formulation, initial matchstick configuration and two solutions of the first Guilford's "match problem" are in Fig. 4.6.

It is important to mentioned that Guilford, in tests evaluations, accepted as point-making only fundamentally different solutions. In other words, three visually different variants of the first solution were not counted.

Formulation, initial matchstick configuration and one solution of the second Guilford's "match problem" are in Fig. 4.7.

According to Guilford, this puzzle can't be solved by persons having cognitive bias "remaining squares must be equal".

In the second part, the Guilford's tests had unusual matchstick puzzles (Guilford and Hoepfner 1971, pp. 425–426). In the first class, for a given matchstick configuration, the task is to get a specified number of squares after removing different numbers of matchsticks. In the second class, for a given matchstick configuration, the task is to get different numbers of squares after removing specified number of matchsticks. As these classes of matchstick puzzles foster development of creative thinking, they should be introduced in mathematics education to help students perform better in posing their own puzzles.

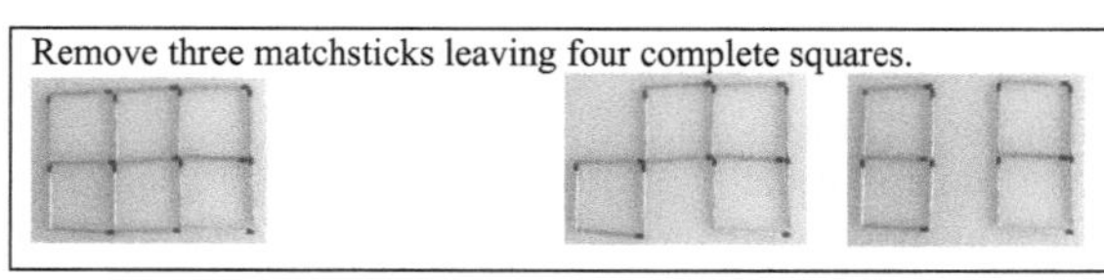

Fig. 4.6 Guilford's "match problem" in which 6 equal squares are converted into 4 equal squares by removing 3 matchsticks (*Source* Author).

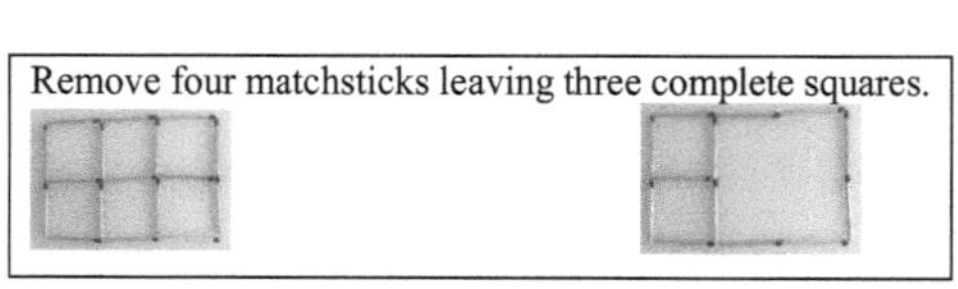

Fig. 4.7 Guilford's "match problem" in which 6 equal squares are converted into 3 unequal squares by removing 4 matchsticks (*Source* Author).

4.4 Difficulty Levels of Arithmetic Matchstick Puzzles with Roman Numerals

In Fig. 4.8 are given initial wrong equations in four arithmetic matchstick puzzles with Roman numerals (A, B, C, and D). Puzzle solvers must find their solutions in form of correct equations (As, Bs, Cs, and Ds) by moving only one matchstick (Knoblich et al. 1999).

Difficulty levels of these puzzles depend theoretically on the constrains that puzzle solvers have self-constructed based on their previous knowledge about ordinary arithmetic operations and on beliefs about possible changes in Roman numerals (Knoblich et al. 1999). The puzzle A is the easiest one, because it only involves "acceptable" changes in two numerals and not in signs "+" and "=". The puzzle D is more difficult because puzzle solvers believe wrongly that Roman numeral X can't be changed as numbers VI or VII. The puzzle B is more difficult than puzzles A and B because puzzle solvers believe that change of the sign "+" into the sign "–" isn't possible. The puzzle C is the most difficult because puzzle solvers consider: the sign "+" can't be converted into the sign "=" and an equation can't have two equality signs. These theoretical predictions were confirmed in experiments by observing the number of solutions found and the times needed to find them (Knoblich et al. 1999).

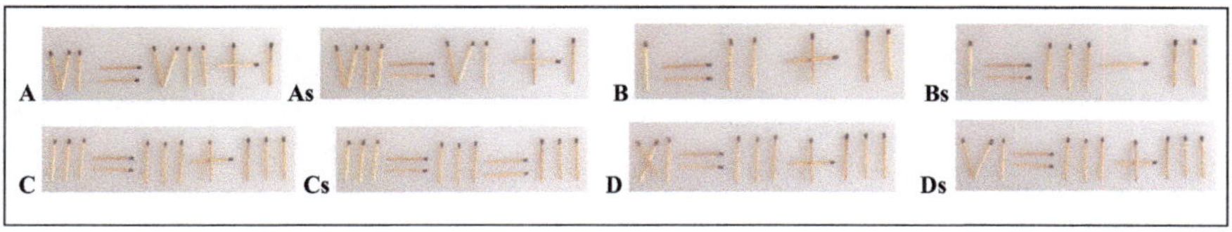

Fig. 4.8 Four arithmetic matchstick puzzles with Roman numerals and their solutions which must be obtained by moving only one matchstick (*Source* Author).

References

Anonymous (1920). Good trick with matches. *The Pathfinder, 27*(1372), 27.

Ballard, J. H. (1915). *Some phases of the psychology of puzzle learning.* Christian Alliance Publishing Company.

Bulbrook, M. E. (1932). An experimental inquiry into the existence and nature of "insight", *The American Journal of Psychology, 44*(3), 410–453.

Guilford, J. P. (1959). Three faces of intellect. *The American Psychologist, 14*(8), 469–479.

Guilford, J. P. (1967*). The nature of human intelligence.* McGraw-Hill Book Company.

Guilford, J. P. (1968). *Intelligence, creativity and their educational implications.* Robert R. Knapp.

Guilford, J. P. & Hoepfner, R. (1971). *The analysis of intelligence.* McGraw-Hill Book Company.

Katona, G. (1940). *Organizing and memorizing. Studies in the psychology of learning and teaching.* Columbia University Press.

Knoblich, G., Ohlsson, S., Haider, H. & Rhenius, D. (1999). Constraint relaxation and chunk decomposition in insight problem solving, *Journal of Experimental Psychology: Learning, Memory, and Cognition, 25*(6), 1534–1555.

Rapport, R. & Yoder, J.A. (2017). *Math lab for kids. Fun, hands-on activities for learning with shapes, puzzles, and games.* Quarry Books.

Reid, J. W. (1951) An Experimental Study of "Analysis of the Goal" in Problem-Solving, *The Journal of General Psychology, 44*(1), 51–69,

Vollmar, P., Haese, M., Haese, R., Haese, S. & Humphries, M. (2008). *Mathematics 6 for the international student.* First edition. Haese and Harris Publications.

Chapter 5
Sophus Tromholt: The Man Who Shaped the World of Matchstick Puzzles

Abstract This chapter presents an informative summary of the big contribution Sophus Tromholt gave to the world of matchstick puzzles. In the Year 1889 Tromholt published book "*Matchstick games. Mental exercise and entertainment*". It was largely dedicated to matchstick puzzles, expanding significantly their forms and number of involved matchsticks. Tromholt shaped the world of matchstick puzzles because numerous of his puzzles were repeated by many posterior authors (Chap. 6). Additionally, in a few puzzles, Tromholt announced after formulation the number of sought different solutions. These solutions were presented in the solution section. From the first to the fifth edition, published in 1892, the number of such puzzle increased from three to twelve. Nevertheless, in many other multiple-solution puzzles, Tromholt published only one solution. Four examples of the "cognitive bias of one solution" are presented for the puzzles that were not repeated by posterior books authors. In their analysis is shown how it is possible to construct additional fundamentally different solutions from a found solution.

5.1 Short Biography of Sophus Tromholt

Sophus Peter Tromholt was born on June 2 of 1851 in Husum, Schleswig, then part of Denmark. His father was customs officer Johan Peter Tromholdt and his mother was Ane Cecilie Keller. Sophus was the first of their five children. The family had interest in natural sciences. Sophus' father published a few articles, one of them on the northern lights.

From 1868 to 1871, Sophus Tromholt was a student at the College of Education in Copenhagen, wishing to become a teacher. During his study, he was mainly interested in meteorology and astronomy. After his graduation, Tromholt worked for eleven years, until 1882, as a teacher at various schools in Aalborg, Horsens, and Bergen. In that period of teaching, he wrote several textbooks on mathematics, geography, and astronomy. Additionally, in 1873 Tromholt published his first scientific paper in German journal "*Wochenschrift für Astronomy, Meteorology und Geography*" about the auroras, he observed during long dark winter nights.

J. Slisko, *Fostering Cognitive Mathematics Skills with Matchstick Puzzles*, Mathematics in Mind, https://doi.org/10.1007/978-3-032-14561-1_5

Fig. 5.1 The photo of Sophus Tromholt (*Source* https://snl.no/Sophus_Tromholt)

In the Year 1883, being 33 years old, Sophus Tromholt (Fig. 5.1) decided to work full-time on the northern lights research.

During the winter 1883–1884, Tromholt was in Iceland, making various observations of the phenomenon. Systematization and analysis of data were carried in Kristiania at the Meteorological Institute, from which Tromholt had a scholarship. His work and contributions to auroral research were presented by Moss and Stauning (2012).

In the Year 1887, being unable to get funding to continue his northern lights research, Tromholt went to live in Germany, first in Dresden and later in Leipzig. To make living, he was a book writer and traveling lecturer on popular science themes, not only in Germany but also in Norway (1890 and 1895) and even in the United States of America (1893). The list of Tromholt's publications is long, containing books and articles on popularization of science, puzzles and mathematical problems for children. Having also a musical talent, he published a few compositions (Horsens Damers Polka and Bergenserinde Polka).

Sophus Tromholt died from pneumonia on May 17 of 1896 at a sanatorium in Thüringen (Germany).

5.2 The Book "*Matchstick Games. Mental Exercise and Entertainment*" and Its Different Editions

In the Year 1889, Tromholt wrote and published, in different languages (German, Danish, and Norwegian), the first edition of the book "*Matchstick games. Mental exercise and entertainment*" (Trolholt 1889). The front page of the book is given in Fig. 5.2.

The book had 123 pages and a modest dimension (11.5 cm × 15 cm). It contained 150 geometric matchstick puzzles and 103 games and amazing physics demonstrations with matchsticks. In the Foreword, Tromholt shown his poetic talent and wrote the following verses:

The matchstick is known as a small, even tiny thing
Usually only noticed when it burns in death.
Yet it modestly contains much entertainment.
Can sharpen the mind and inspire cheerful play.

Fig. 5.2 Front page of the first edition of the book "*Matchstick games. Mental exercise and entertainment*" (*Source* Photo by Heinrich Hemme)

> …
> With what I learned from others and what I myself invented
> May the little book now wander cheerfully into the wide world
> And seek friends where games and jokes are enjoyed.
> Every happy hour it can provide
> will be a happy hour for the one who invented it.

The number of 150 matchstick puzzles in Tromholt's book was in strong contrast with numbers of puzzles in previous books. Braun's book (Braun 1876) had 8 matchstick puzzles while in Mittenzwey's book their number was 19. It means that Tromholt made an enormous creative effort to expand the collection of possible matchstick puzzles to its greater extension. What Tromholt "learned from others" were only a few puzzles, "borrowed" from Braun and Mittenzwey. All others were the matchstick puzzles he invented. They shaped the world of matchstick puzzles from then on. Namely, as it will be shown in Chap. 6, Tromholt's puzzles and solutions were repeated by many posterior book authors.

The Tromholt's book was sold very well and had second and third unchanged editions in the same year. In the Year 1890 comes out the fourth edition (Tromholt 1890) with some changes (126 pages and 158 matchstick puzzles). The fifth edition, largely improved (Tromholt 1892), was the last one Tromholt involved with. It had 150 pages and 162 matchstick puzzles. All posterior editions were the same. Between 1892 and 1915, *twelve additional editions* were published. This is an impressive success that no other book on matchstick puzzles has repeated until today!

5.3 New Classes of Matchstick Puzzles in Tromholt's Book

Sophus Tromholt invented various new classes of matchstick puzzles. Being this a short critical review, in what follows come only two examples of those classes.

In the first class there are many puzzles in which the shapes of initial and final configurations of matchsticks were very different, making their solving very hard and, consequently, likely less attractive for puzzle solvers. This class can be divided into two subclasses.

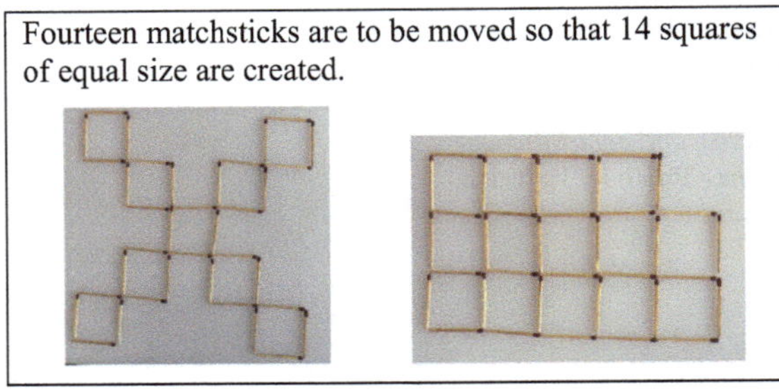

Fig. 5.3 One example of puzzle with very different initial and final matchstick configurations, both made from 36 matchsticks (*Source* Author)

In the first subclass, the initial number of matchsticks is relatively small. The formulation, initial matchstick formulation, and solution of one of such examples with 36 matchsticks (Tromholt 1889, p. 22, p. 106) are given in Fig. 5.3.

To the second subclass belong the puzzles in which the number of matchsticks in initial configuration is relatively big. Although there were puzzles with bigger initial number of matchsticks, the example is a puzzle that started with 60 matchsticks. Its formulation, initial configuration, and solution (Tromholt 1889, p. 25, p. 108) are given in Fig. 5.4.

The second new class of puzzles were those in which the initial configuration contains matchsticks with free ends. In the first subclass, various matchsticks had one free end. The formulation, initial matchstick configuration, and solution (Tromholt 1889, p. 64, p. 146) of one example are presented in Fig. 5.5.

In the second subclass are those puzzles in which initial configuration is a chain of matchsticks where only the first and last matchstick have a free end (Tromholt 1889, p. 79, p. 118). One example is given in Fig. 5.6.

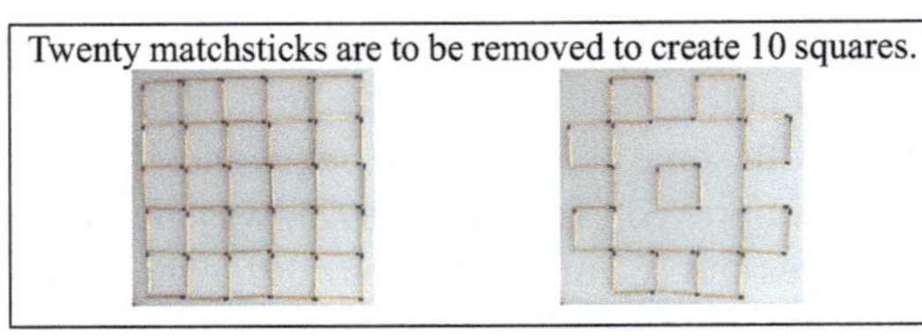

Fig. 5.4 One example of puzzle with very different initial and final matchstick configurations, the first with 60 and the second with 40 matchsticks (*Source* Author)

Fig. 5.5 One example of puzzle in which the initial configuration contains matchsticks with free ends (*Source* Author)

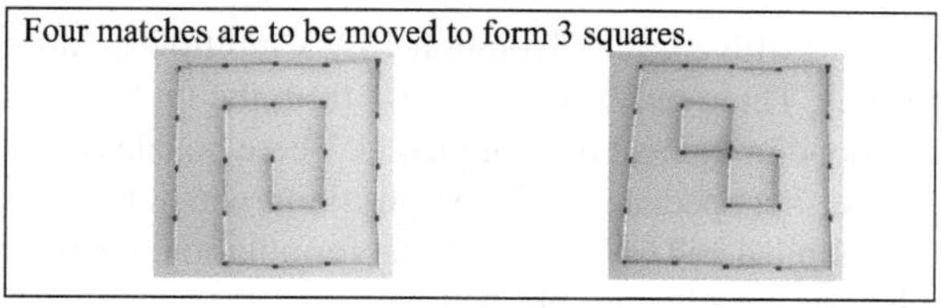

Fig. 5.6 One example of puzzle in which the initial configuration contains only two matchsticks with free ends (*Source* Author)

5.4 Matchstick Puzzles with Announced Multiple Solutions

Tromholt introduced another important feature to the world of matchstick puzzles: announcement and publication of the number of possible different solutions readers are supposed to find. A critical analysis of the realization of such an important idea shows that it wasn't developed in a totally satisfactory manner.

In the first edition of the book "*Matchstick games. Mental exercise and entertainment*" (Tromholt 1889), multiple solutions were mentioned as a part of formulation only in three matchstick puzzles (Tromholt 1889, p. 33, p. 110; p. 40, p. 112, p. 66, p. 114). One worth of comment is the first of them. Puzzle's formulation, initial matchstick configuration, and 3 solutions (Tromholt 1889, p. 33, p. 110) are in Fig. 5.7.

The third solution is an example of cognitive bias "uncritical thinking". Namely, the solution is wrong because it contains not two but three squares!

In the fourth edition (Tromholt 1890), multiple solutions were mentioned only in four puzzles. Three puzzles were those from the first edition, and in one puzzle from first edition a second solution was added (Tromholt 1889, p. 6, p. 107). Its formulation, initial matchstick configuration, and solutions are presented in Fig. 5.8.

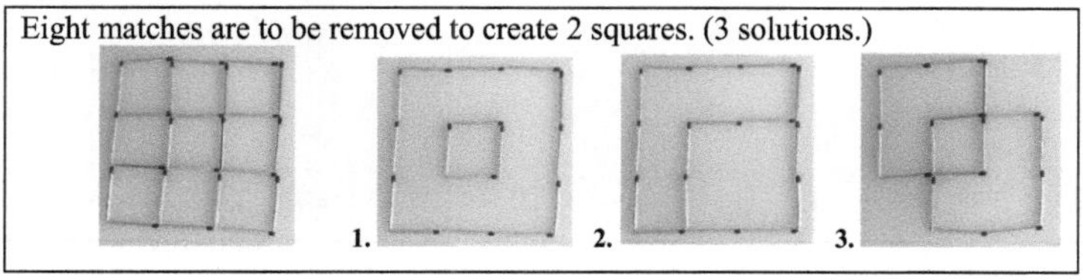

Fig. 5.7 The puzzle with three solutions (*Source* Author)

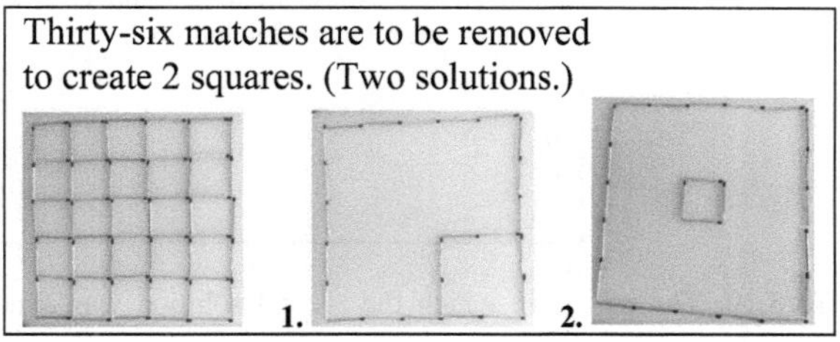

Fig. 5.8 A new puzzle with two solutions in the fourth edition (*Source* Author)

In the fifth edition (Tromholt 1892), multiple solutions were mentioned in twelve puzzles. Three puzzles were those from the first and fourth editions. Let's see what was done in the rest of nine puzzle. Three puzzles are special cases.

The first special case is the puzzle in which the added third solution was wrong (Tromhollt 1892, p. 13, p.128). Its formulation, initial matchstick configuration, and solutions are given in Fig. 5.9.

The error in the third solution is again that the solution contains not 2 but 3 squares!

The second special case is the puzzle (Tromholt 1892, p. 13, p. 128) in which two solutions were announced but only one, from two previous editions, was published! Its formulation, initial matchstick configuration, and solutions are presented in Fig. 5.10.

The third special case is the puzzle (Tromholt 1892, p. 17, p. 129) in which two published solutions have been already published separately as single solutions in previous editions. Its formulation, initial matchstick configuration, and solutions are presented in Fig. 5.11.

The first solution is single solution published in the fourth edition (Tromholt 1890, p. 10, p. 108), while the second solution is single solution published in the first edition (Tromholt 1889, p. 11, p. 108). The third special case shows clearly that Tromholt changed his mind twice while thinking about the solution of this puzzle!

In remaining six puzzles, Tromholt presented correctly the announced multiple solutions.

In two of them (Tromholt 1892, p.11, p. 127; Tromholt 1892, p. 50, p. 138) only one new solution was added to the single one already published in previous editions. Their

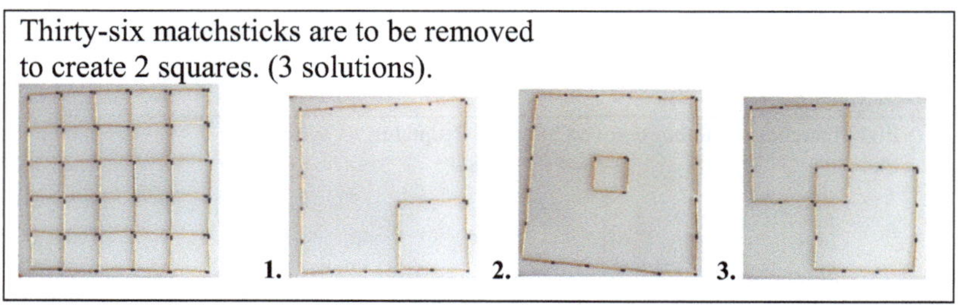

Fig. 5.9 Another puzzle in which the third solution is erroneous (*Source* Author)

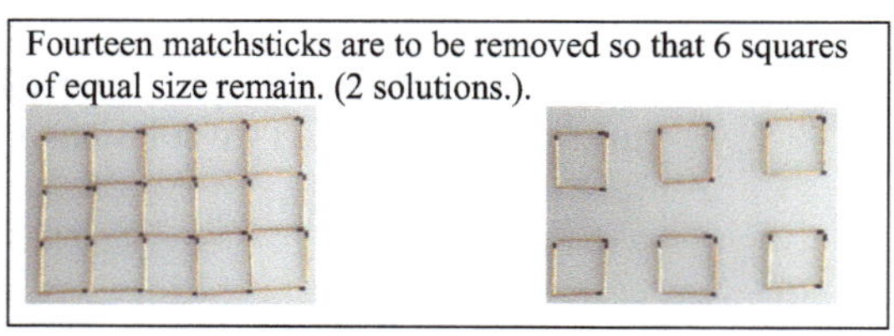

Fig. 5.10 The puzzle in which the second solution was announced but not published (*Source* Author)

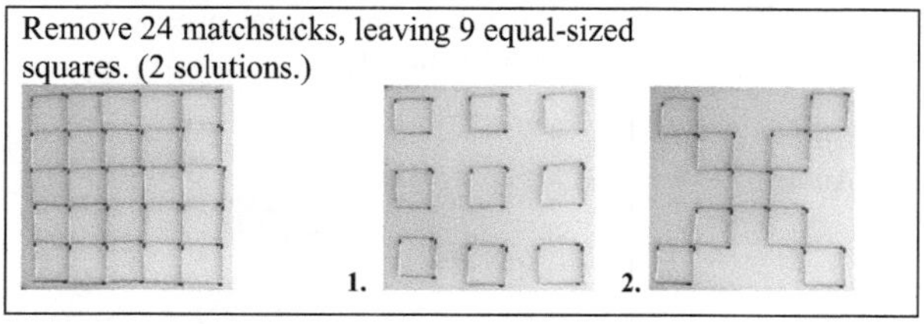

Fig. 5.11 The puzzle in which two published solutions have been published separately as single solutions in previous editions (*Source* Author)

formulations, initial matchstick configurations, and solutions are given in Figs. 5.12 and 5.13.

In two puzzles (Tromholt 1892, p. 30, p. 133; Tromholt 1892, p. 49, p. 137) two new solutions were added to the single one already published in previous editions. Their formulations, initial matchstick configurations, and solutions are given in Figs. 5.14 and 5.15.

In one puzzle (Tromholt 1892, p. 19, p. 139) four new solutions were added to the single one already published in previous editions. Its formulations, initial matchstick configurations, and 5 solutions are given in Fig. 5.16.

Finally, in the fifth edition there was a new puzzle with three announced and published solutions (Tromholt 1892, p. 110, p. 147). Its formulations, initial matchstick configurations, and solutions are given in Fig. 5.17.

It is important to note that, in majority of examples, Tromholt's single published solution in the first and fourth editions was symmetric, while added solutions were

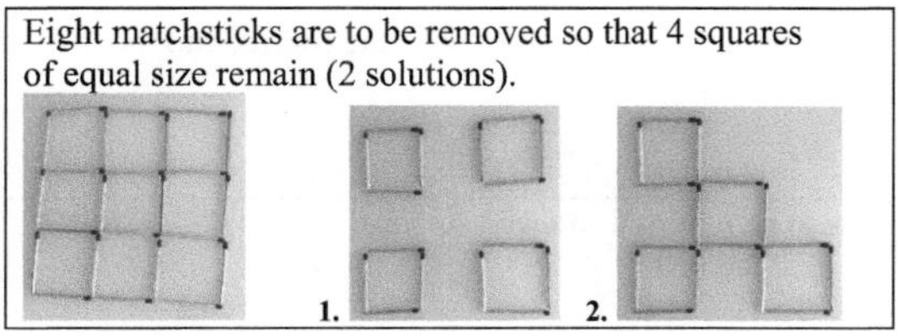

Fig. 5.12 The first example of the puzzle to which the second solution is added in the fifth edition (*Source* Author)

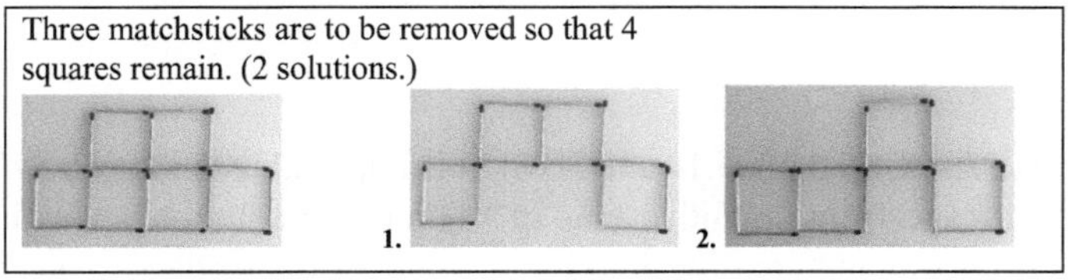

Fig. 5.13 The second example of the puzzle to which the second solution is added in the fifth edition (*Source* Author)

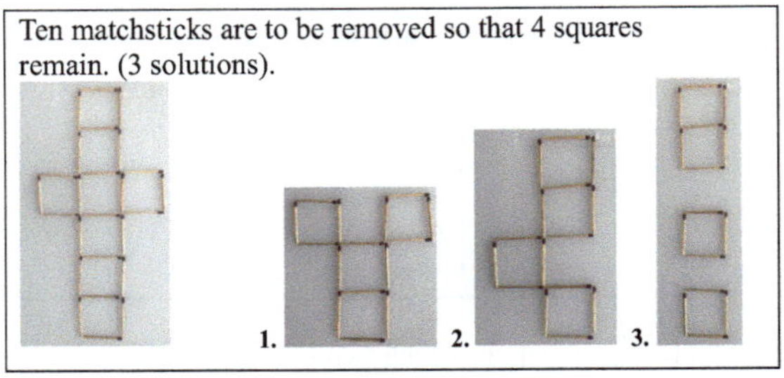

Fig. 5.14 The first example of the puzzle to which the second and the third solutions were added in the fifth edition (*Source* Author)

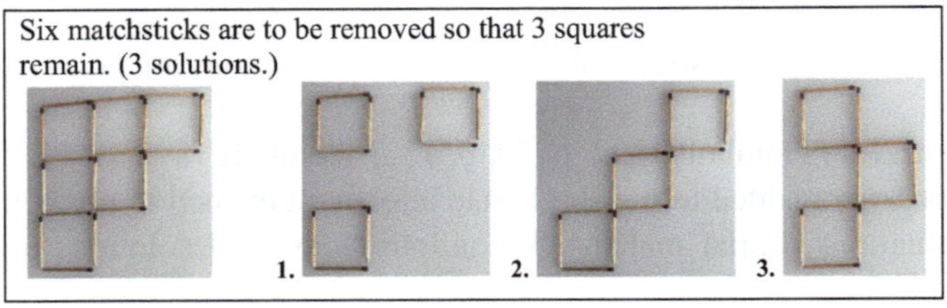

Fig. 5.15 The second example of the puzzle to which the second and the third solutions were added in the fifth edition (*Source* Author)

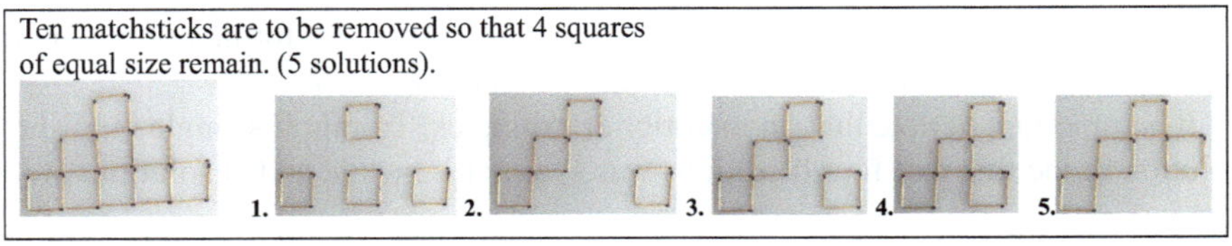

Fig. 5.16 The puzzle to which the second, the third, the fourth, and the fifth solutions were added in the fifth edition (*Source* Author)

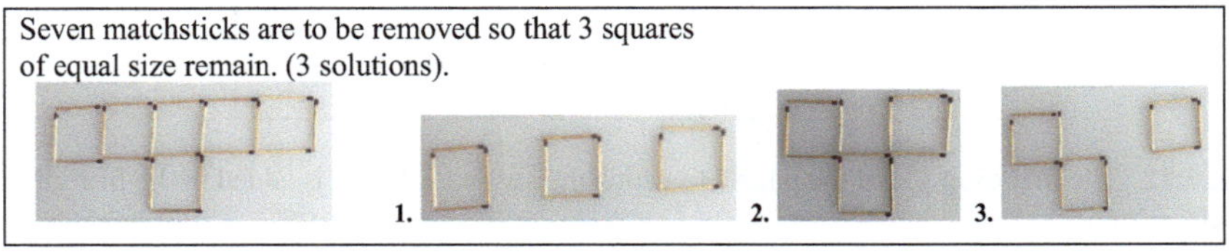

Fig. 5.17 New puzzle with three announced and published solutions in the fifth edition (*Source* Author)

asymmetric (Figs. 5.12, 5.13, 5.14 and 5.16). It is impossible to know how Tromholt was able to escape from cognitive bias of "symmetric solution". Was it a posterior strike of critical and creative thinking or some of the readers wrote him about additional asymmetric solutions?

5.5 Tromholt's Multiple-Solution Matchstick Puzzles with Only One Solution Published

As it was seen, Tromholt had interest in announcing and publishing multiple solutions to his puzzles. Nevertheless, in many puzzles he failed to find more than one solution. In this section four examples of such puzzles are presented and analyzed. In three of them, single published solution is again symmetric!

These puzzles are among unpopular Tromholt's puzzles because they weren't repeated by any of posterior book authors.

1. Remove 6 matchsticks to get 10 equal-sized squares

The first example is a puzzle published in the same form in all three editions (Tromholt 1889, p. 44, p. 113; Tromholt 1890, p. 43, p. 117; Tromholt 1892, p. 53, p.139). Its formulation, initial matchstick configuration, and solution are in Fig. 5.18.

It is easy to notice that there are additional fundamentally different solutions which are asymmetric. Katona's strategy for getting different configurations with the same number of squares is useful. In this case, it means that by moving matchsticks two "old" squares should be eliminated and two "new squares" should be created.

These solutions can be obtained by two ways. In the first way, one puts in different places two matchsticks from the first square in the second row. In the second way, one puts in different places two matchsticks from the third square in the second row. Three asymmetric solutions that can be obtained by the first way are presented in Fig. 5.19.

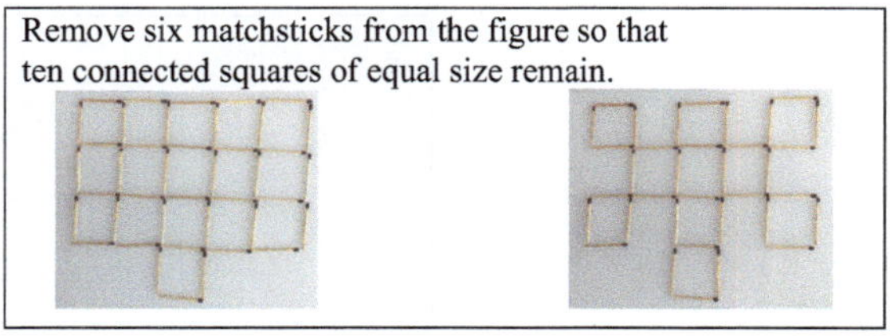

Fig. 5.18 Formulation, initial matchstick configuration, and symmetric solution of the puzzle in the first example (*Source* Author)

Fig. 5.19 Three additional *asymmetric solutions* for the puzzle in the first example (*Source* Author)

2. Remove 21 matchsticks to get 4 squares

The second example is the puzzle published in the same form in all three editions (Tromholt 1889, p. 29, p. 110; Tromholt 1890, p. 28, p. 114; Tromholt, 1892, p. 37, p. 135). Its formulation, initial matchstick configuration, and solution are in Fig. 5.20.

There is another fundamentally different solution with two 2 × 2 squares and two 1 × 1 squares (Fig. 5.21).

As nothing was said about the forms of 4 sought squares, Braun's strategy helps to find additional fundamentally different solutions. Namely, after removing 21 matchsticks from 45 initial matchsticks, 24 matchsticks are left for making 4 squares.

The first possibility is to form one 3 × 3 square (12 matchsticks) and three 1 × 1 squares (12 matchsticks). Two different realizations of this possibility, in which three 1 × 1 squares lay outside of the 3 × 2 square, are presented in Fig. 5.22.

Each one of three outside 1 × 1 squares can be placed into central position inside the 3 × 3 square. Doing it for three 1 × 1 squares in the first solution in Fig. 5.22, one

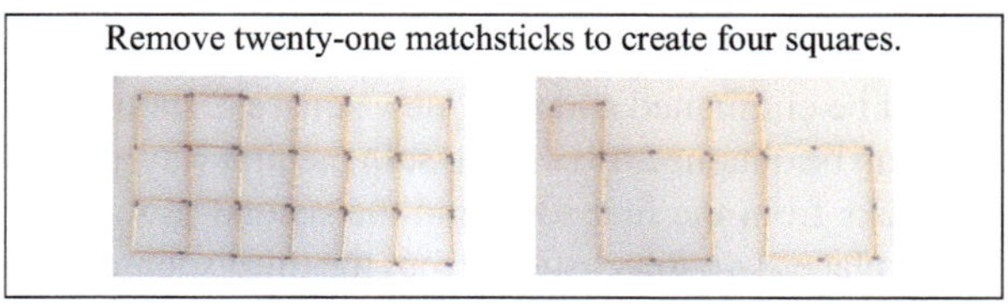

Fig. 5.20 Formulation, initial matchstick configuration, and solution of the puzzle in the second example (*Source* Author)

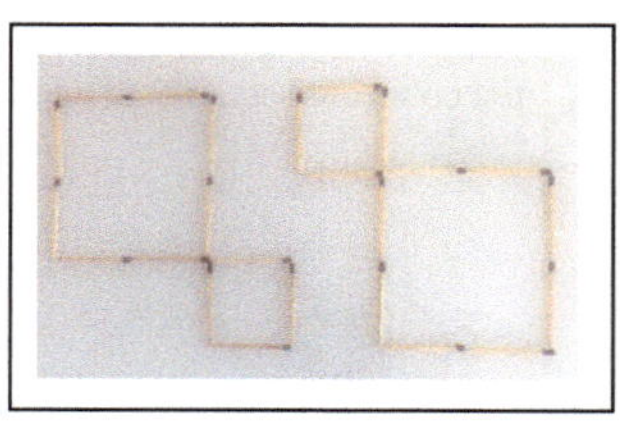

Fig. 5.21 Additional solutions for the puzzle in the second example (*Source* Author)

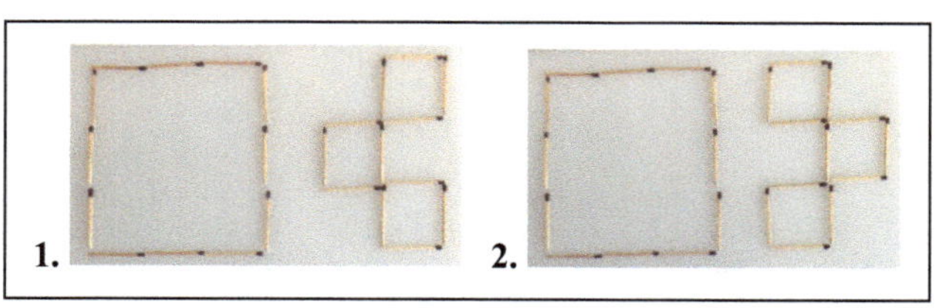

Fig. 5.22 Two additional solutions for the puzzle in the second example with three 1 × 1 squares laying outside the 3 × 3 square (*Source* Author)

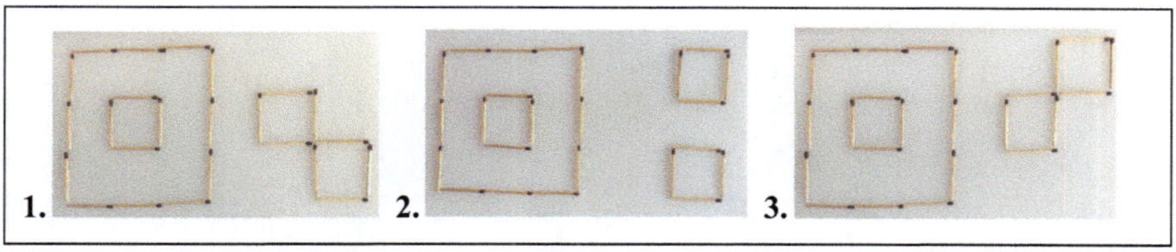

Fig. 5.23 Three additional solutions for the puzzle in the second example with one 1 × 1 square inside 3 × 3 square (*Source* Author)

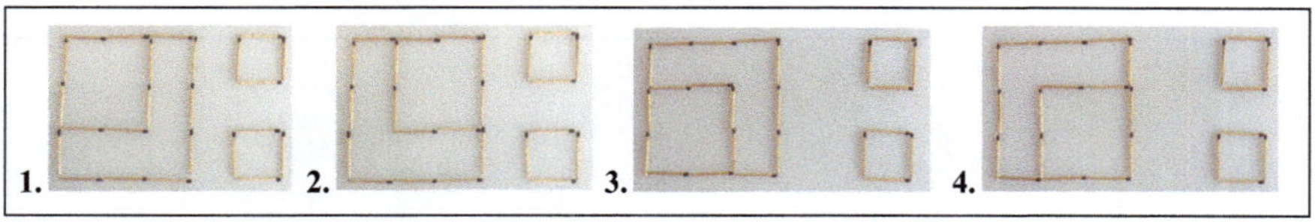

Fig. 5.24 Four additional asymmetric solutions for the puzzle in the second example (*Source* Author)

gets three visually different solutions, being first two also fundamentally different (Fig. 5.23).

Additionally, the matchsticks from each one of outside 1 × 1 squares can make one 2 × 2 square inside the 3 × 3 square at four different positions. Doing it for the middle outside 1 × 1 square in the second solution in Fig. 5.22, one gets four visually different solutions presented in Fig. 5.24.

Remove 24 matchsticks to get 3 squares

The third example is the puzzle published in the same form in all three editions (Tromholt 1889, p. 43, p. 113; Tromholt 1890, p. 42, p. 117; Tromholt 1892, p. 52, p. 139). Its formulation, initial matchstick configuration, and symmetric solution are in Fig. 5.25.

It is easy to notice that central 1 × 1square can become one 2 × 2 square, "borrowing" four matchsticks from the 3 × 3 square. This can be carried out in four ways, getting four visually different asymmetric solutions (Fig. 5.26).

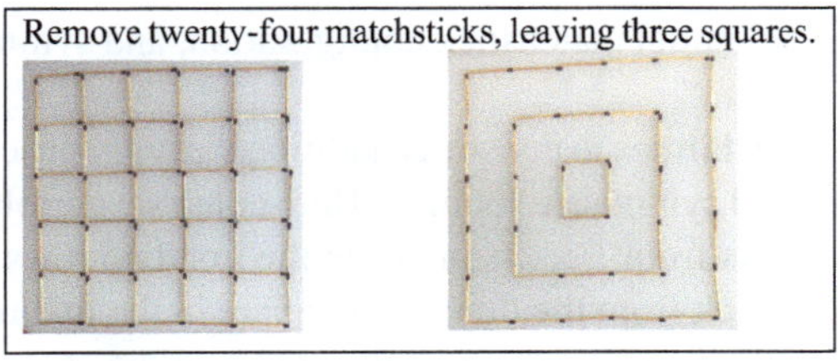

Fig. 5.25 Formulation, initial matchstick configuration, and symmetric solution of the puzzle in the third example (*Source* Author)

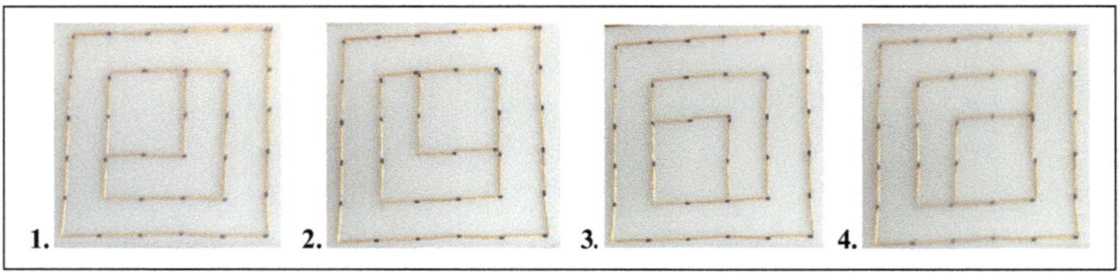

Fig. 5.26 Four additional asymmetric solutions with new 2 × 2 square inside 3 × 2 square for the puzzle in the third example (*Source* Author)

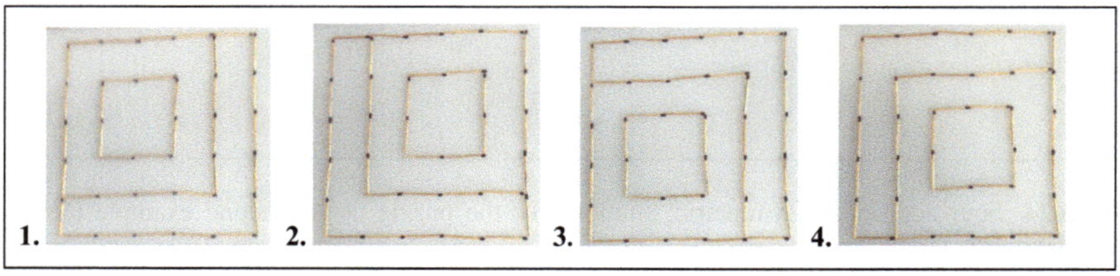

Fig. 5.27 Four additional asymmetric solutions with new 4 × 4 square and 2 × 2 square inside 5 × 5 square for the puzzle in the third example (*Source* Author)

Using Brown's strategy, one can get new additional solutions for the puzzle in the third example.

When 24 matchsticks are removed from 60 initial matchsticks, one is left with 36 matchsticks for forming 3 different-sized squares. A possible solution is formed from one 5 × 5 square (20 matchsticks used), having inside one 4 × 4 square (8 additional matchsticks used) and one 2 × 2 square (8 additional matchsticks used). It has four visually different variants (Fig. 5.27).

Tromholt's symmetric solution would be unique possible solution only in a restricted puzzle reformulation in which remaining squares must be non-touching squares.

Remove 16 matchsticks to get 1 large and 9 small squares

The fourth example is the puzzle published in the same form in all three editions (Tromholt 1889; p. 40, p. 112; Tromholt 1890, p. 38, p. 115; Tromholt 1892, p. 48, p.137). Its formulation, initial matchstick configuration, and symmetric solution are given in Fig. 5.28.

There are many possibilities to construct additional asymmetric solutions making changes in the published symmetric solution. Here come seven solutions one can get by replacing matchsticks from 1 × 1 squares in the first two rows.

Replacing 1 matchstick from the first 1 × 1 square in the first row, it can become "new" first square in the first row or the only square in the second row. Replacing 2 matchsticks from the second 1 × 1 square in the first row, it can become the only square in the second row. Replacing 1 matchstick from the third 1 × 1 square in the

Take away sixteen matchsticks, creating one large square and nine small squares.

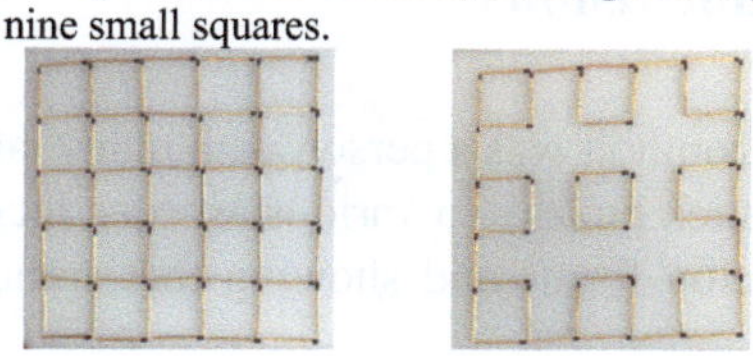

Fig. 5.28 Formulation, initial matchstick configuration, and symmetric solution of the puzzle in the fourth example (*Source* Author)

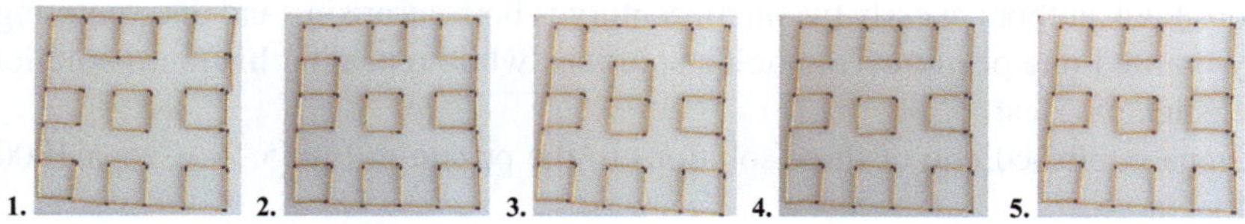

Fig. 5.29 Five asymmetric solutions obtained by replacing matchsticks from the three 1 × 1 squares in the first row (*Source* Author)

first row it can become "new" third square in the first row or the only square in the second row. These five possibilities are presented in Fig. 5.29.

Two additional solutions are obtained in this way: Replacing 2 matchsticks from the first 1 × 1 square in the second raw, it becomes "new" first square in the second row. Replacing 2 matchsticks from the third square in the second row, it becomes "new" third square in the second row (Fig. 5.30).

Additional 5 visually different asymmetric solutions can be obtained by carrying out with matchsticks in the third row the changes like those done in the first row.

Tromholt's symmetric solution would be unique solution only in a restricted puzzle reformulation in which is prohibited that nine inside 1 × 1 squares share one matchstick.

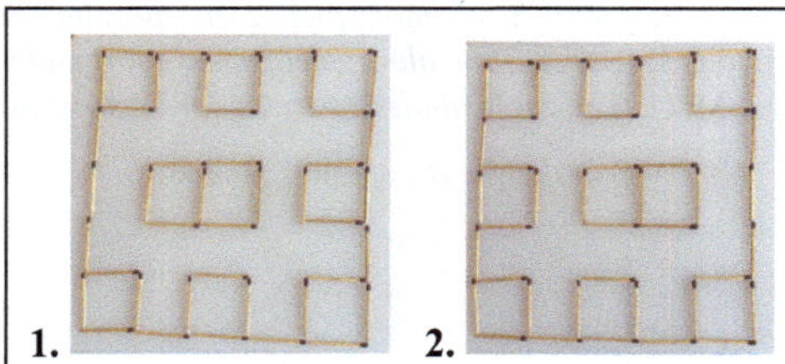

Fig. 5.30 Two asymmetric solutions obtained by replacing 2 matchsticks from the first and the third square in the second row (*Source* Author)

5.6 Conclusions

Sophus Tromholt was a person with many talents who shaped profoundly the world of matchstick puzzles in various aspects, increasing greatly their number and initial configuration forms and showing that some of matchstick puzzles have multiple solutions.

Considering as very important this last contribution of Tromholt for fostering creative thinking of book readers in solving matchstick puzzles, its treatment by various book authors in early twentieth century is both surprising and disappointing. They omitted one or more Tromholt's solutions while repeating his multi-solution matchstick puzzles!

Czepa published one of three solutions to the puzzle in Fig. 5.15 (Czepa 1900, p. 58).

Wesel published one of two solutions to the puzzle in Fig. 5.13 (Wesel 1912, p. 41), one of three solutions to the puzzle in Fig. 5.14 (Wesel 1912, p. 69), and two of three solutions to the puzzle in Fig. 5.17 (Wesel 1912, p. 30).

Lüders published only one of five solutions to the puzzle in Fig. 5.16 (Lüders 1919, p. 18)!

References

Braun, F. (1876). *Der junge Mathematiker und Naturforscher*. Verlag von Otto Spamer.

Czepa, A. (1910). *Streichholz-Spielereien: Kunststücke und Scherze nebst Auflösungen.* Otto Maier Verlag.

Lüders, C. E. (1919). *Der Streichholzkünstler in der Westentasche.* Verlag von G. Danner.

Mittenzwey, L. (1880). *Mathematische Kurzweil.* Julius Klinkhardt.

Moss, K., & Stauning, P. (2012). Sophus Peter Tromholt: an outstanding pioneer in auroral research. *History of Geo-and Space Sciences*, *3*(1), 53–72.

Tromholt, S. (1889). *Streichholzspiele. Denksport und Kurzweil.* First edition. Otto Spamer Verlag.

Tromholt, S. (1890). *Streichholzspiele. Denksport und Kurzweil* Fourth edition. Otto Spamer Verlag.

Tromholt, S. (1892). *Streichholzspiele. Denksport und Kurzweil.* Fifth edition. Otto Spamer Verlag.

Wesel, M. (1912). *Streichholzscherze in der Westentasche.* Verlag G. Danner.

Chapter 6
Tromholt's Matchstick Puzzles Published by Posterior Book Authors

Abstract This chapter considers critically formulations and solutions of some Tromholt's matchstick puzzles that were later published by two or more authors. It is an attempt to discover and describe what might be an important feature of "collective culture" of matchstick puzzle posing and solving. Special attention is given to two different approaches posterior book authors had to Tromhol's solutions. In the first case, some posterior book authors published the solutions that were visually different from Tromholt's solutions. In the second, more important case, for matchstick puzzles with multiple, fundamentally different solutions, almost all posterior authors just repeated Tromholt's solutions. In the worst case, posterior authors published a puzzle with erroneous solution that Tromholt himself has excluded in later editions. Additionally, the absence of these numerous additional, fundamentally different solutions is an initial evidence that posterior book authors on matchstick puzzles didn't have enough *creativity inclination* that is necessary to find all possible solutions of the puzzles they deal with. Omitting those solutions, they don't give book readers deserved opportunities to practice and improve their creativity and visual intelligence.

6.1 Visually Different Solutions to Matchstick Puzzles: A Closer Look

As it was mentioned before, the first matchstick puzzle had two visually different solutions. The same situation occurs with many other matchstick puzzles that might have even more than two visually different solutions.

Nevertheless, starting with Tromholt, book authors commonly don't warn book readers and supposed puzzle solvers that their published single solution isn't unique and that are other visually different solutions that might be obtained by rotation or mirror reflection from the published one.

J. Slisko, *Fostering Cognitive Mathematics Skills with Matchstick Puzzles*, Mathematics in Mind, https://doi.org/10.1007/978-3-032-14561-1_6

In revision of a big number of puzzle books, I found only two authors who inform the readers about this important fact related to different visual versions of a solution. The first book is a standard book on matchstick puzzle in which one can read (Anonymous 2017, p. 7):

> Many of these puzzles may have more than one possible solution, including reflections and rotations of those given at the back of this book; space does not permit us to show more than one solution to each of the puzzle, but at least if you get stuck, you will be able to fin an answer.

The second book isn't adequate for educational aims. It uses unpolite language and "loo brushes" instead of matchsticks. Nevertheless, the important message for its readers is (Roffy 2014, p. 79):

> Your answer could be a mirror image or a rotated version – it's still the same answer.
>
> If you found others, then well done, you can tell your friends that you are king of the loo brush!

Without the warning about multiplicity of visually different solutions, the readers who get solutions which are different from the author's one might be undeservedly led to a wrong conclusion that their solutions are wrong, losing confidence in their creative skills! The case of one real but anonymous reader is documented and discussed in Chap. 9.

The probabilities that different puzzle solvers could find visually different solutions to a puzzle can't be reduced to zero. It seems intuitively correct to expect that in a sufficiently big group of puzzle solvers all possible visually different solutions will appear, although with different percentages.

This expectation will be explored by looking at different solutions to three popular Tromholt's puzzles published by posteriors book authors.

6.2 Move 2 Matchsticks to Get 5 Equal Squares

The first popular puzzle has four visually different solutions, but Tromholt published only one (Tromholt 1889, p. 2, p. 101; Tromholt 1890, p. 1, p. 106; Tromholt 1892, p. 7, p. 127). Its formulation, initial matchstick configuration, and solution are presented in Fig. 6.1.

Thirteen posterior book authors repeated this Tromholt's puzzle. Twelve of them, as Tromholt, published only one solution. Only one author (Anonymous 1996, p. 8, p. 82, p. 133, p. 170) published two visually different solutions. Five of posterior book authors repeated Tromholt's solution, too (Lüders 1919, p. 40; Paraquin 1970, p. 2, p. 26; Hansell 1981, p. 49, p. 74; Anonumous, 1996, p. 8, p. 133; Picon 2002, p. 20, p. 115).

Three additional solutions which are visually different from the solution of Tromholt and mentioned posterior book authors are presented in Fig. 6.2.

Fig. 6.1 Formulation, initial matchstick configuration, and a single solution of the first popular Tromholt's puzzle (*Source* Author)

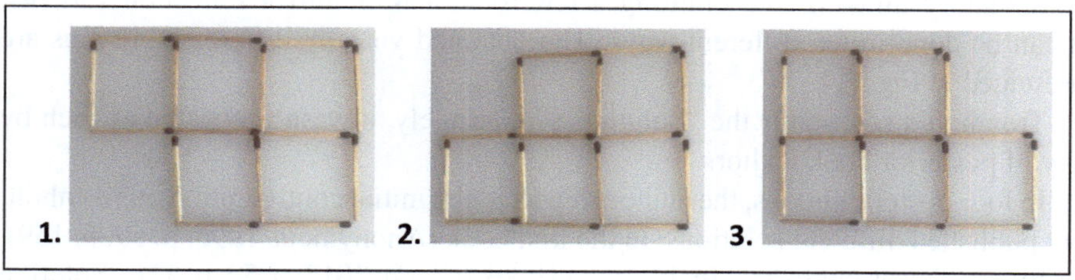

Fig. 6.2 Three additional solutions which are visually different from Tromholt's solution (*Source* Author)

The first additional solution, visually different from Tromholt's solution, was published by three posterior book authors (Van Delft and Botermans 1978, p. 189; Anonymous 2014, p. 106; Anonymous 2017, p. 112).

The second additional solution, visually different from Tromholt's solution, was published by two posterior book authors (Anonymous 1996, p. 170; Clarke 2017, p. 63).

The third additional solution, visually different from Tromholt's solution, was published by four posterior book authors (Czepa 1910, p. 61; Botermans 2006, p. 191; Grund-Thorpe 2006, p. 61; Colson 2012, p. 187).

Although this sample contains only posterior books, the percentages of different solutions don't differ greatly. The biggest percentage (38%) belongs to Tromholt's solution. The next percentage (30%) belongs to the third additional solution. The percentage (23%) belongs to the first additional solution. The smallest percentage (15%) belongs to the second additional solution.

One can wonder why solutions in which single small square is on the right side of four small squares (Tromholt's solution and third additional solutions) are more frequent (68%) than the solutions in which single small square is on the left side of four small squares (38%). These percentages don's sum 100% because one book author published two different solutions.

6.3 Move 2 Matchsticks to Get 7 Equal-Sized Squares

The second popular puzzle has ten visually different solutions, but Tromholt published only one (Tromholt 1889, p. 3, p. 101; Tromholt 1890, p. 3, p. 106; Tromholt 1892, p. 10, p. 127). Its formulation, initial matchstick configuration, and solution are presented in Fig. 6.3.

Katona's strategy helps to find all visually different solutions. To make 7 squares from 8 squares, one has, by moving 2 matchsticks, to "destroy" 2 "old squares" and make 1 new square at the only possible position (upper left vertex). Using combinatory thinking and omitting a few rotational or mirror reflection variants, it can be done in ten different ways. The obtained visually different solutions are presented in Fig. 6.4.

Tromholt's solution is the "Solution 8". Strangely, it wasn't repeated as such by any of posterior book authors.

In four posterior books, the authors repeated the initial configuration of Tromholt, but published different solutions. In the first book (Anonymous 1996, p. 59, p. 159), "Solution 3" was published. In the second book (Clarke 2017, p. 52, p. 52), "Solution 5" was published. In the third book (Anonymous 2014, p. 69, p. 119), "Solution 6"

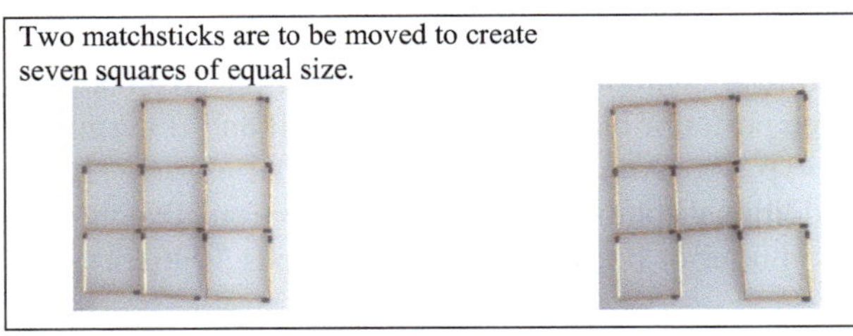

Fig. 6.3 Formulation, initial matchstick configuration and a single solution of the second popular Tromholt's puzzle (*Source* Author)

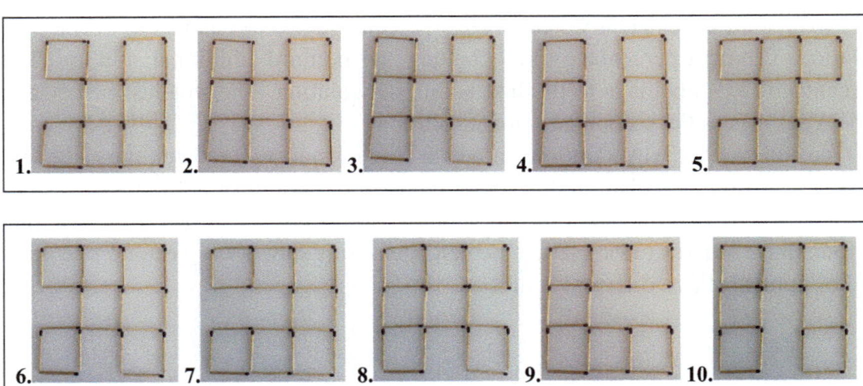

Fig. 6.4 Ten visually different solutions to the second popular Tromholt's puzzle (*Source* Author)

was published. In the fourth book (Paraquin 1970, p. 5, p. 27), "Solution 5" and "Solution 6" were published.

In two posterior books, authors changed Tromholt's initial matchstick configuration (Czepa 1910, p. 42, p. 61; Botermans 2006, p. 55, p. 195). In the first book, "Solution 6" was published, while in the second book a mirror image of the "Solution 6" was published.

The number of possible visually different solutions is bigger than 10 and the number of posterior book authors is 6. As only one author (Paraquin 1970) published 2 different solutions, various visually different solution didn't appear in any of considered books. It is worth noting, that two published solutions ("Solution 3" and "Solution 5") differ fundamentally from Tromholt's solution ("Solution 8"). In addition, the second fundamentally different solution ("Solution 4") wasn't published either by Tromholt or by any of posterior book authors.

6.4 Move Some Matchsticks to Get 7 Equal-Sized Squares

The third popular matchstick puzzle has 8 visually different solutions, but Tromholt published only one (Tromholt 1889, p. 54, p. 114; Tromholt 1890, p. 53, p. 117; Tromholt, 1892, p. 63, p. 139). Its formulation, initial matchstick configuration, and solution are presented in Fig. 6.5.

In three posterior books, the authors affirm that the puzzle "move 3 matchsticks to get 7 equal-sized squares" has eight visually different solutions. Nevertheless, their approaches differ substantially.

In the first book (Johnson-Laird 2005, p. 10), the existence of "eight solutions" is only a verbal fact.

In the second book (Orleans and Orleans 1994, p. 13, p. 250), the authors don't present any visualization. Instead, they verbally describe how to get "eight solutions":

> Move 3 sticks from any square and use to complete an adjacent square and 2 opposite squares.

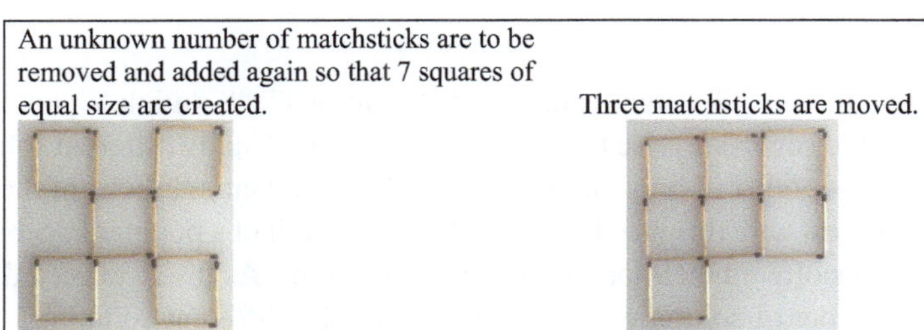

Fig. 6.5 Formulation, initial matchstick configuration, and a single solution of the third popular Tromholt's puzzle (*Source* Author)

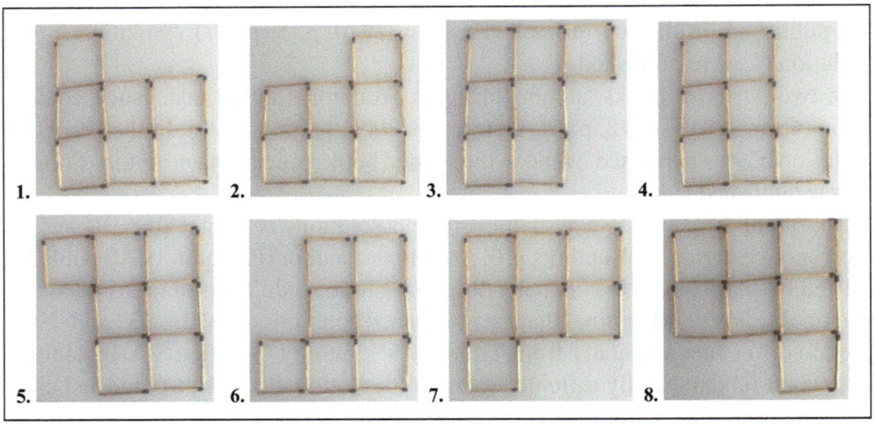

Fig. 6.6 Eight visually different solutions to the third popular Tromholt's puzzle (*Source* Author)

In the third book (Gutrie and Preston 2017, p. 208, p. 254), visualization of one solution is presented (see below "Solution 6"), along with the procedure how to get eight visually different solutions:

> Move three from any corner square and complete the square next door and two the opposite squares.

Eight visually different solutions are presented in Fig. 6.6.

"Solution 1" was published in two books (Seymoure 1991, p. 119; Huang et all, 2020, p. 20). In the first book, possibility of other solutions is mentioned, without giving a specific number. The other authors didn't mention the existence of additional, visually different solutions, and published only one solution. These authors published the following solutions:

"Solution 2" (Uhl 1947, p. 43);

"Solution 3" (Czepa 1910, p. 57);

"Solution 7", Tromholt's solution, (Wesel 1912, p. 48; Williams 2013; p. 111; Anonymous 2014, p. 101).

"Solution 8" (Paraquin 1970, p. 26).

Nobody of these book authors published "Solution 4", "Solution 5", or "Solution 6". This absence isn't a surprise because the number of visually different solutions is bigger than the number of book authors who published their solutions of the puzzle.

When a contest with this puzzle was published on wall of a popular Facebook site, 146 participants offered their individual single solutions. As it was expected, among such a big number of solutions were all eight visually different solutions. "Collective creativity" of 146 participants was much bigger than creativity of "collective creativity" presented by 7 book authors! The percentages of different solutions will be exposed and discussed in Chap. 10.

6.5 Tromholt's Puzzles and Solutions Repeated by Two or More Posterior Book Authors: In Search of Fundamentally Different Solutions

The aim of this chapter is to show that features of Tromholt's approaches to solving matchstick puzzles were shared by posterior book authors. By repeating Tromholt's puzzles and solutions, they didn't try to reshape the field of matchstick puzzles solving making it more suitable for development of readers' creativity, visual intelligence and other cognitive mathematics skills.

Instead of it, posterior authors revealed various cognitive biases, like "one puzzle—one solution", "solution must be symmetric", "sought squares must be of equal size", or "formulation change doesn't matter". These cognitive biases are most important features of "collective culture" of book authors about matchstick puzzles.

Selected puzzles that follow are only initial illustrative evidence of the presence of these "cognitive biases" among various posterior book authors. This evidence could be easily extended if this book wouldn't have restricted number of pages.

The absence of visually different solutions could be "justified" by the (unstated) argument that they are only rotation or mirror versions of published solutions. The absence of fundamentally different solutions is strong evidence of the presence of cognitive bias shared by the book authors. To show how is possible to escape from these cognitive biases, some puzzle solving strategies are presented to help find fundamentally different solutions.

6.6 Remove 18 Matchstick to Get 3 Squares

The first of selected puzzles (Tromholt 1889, p. 20, p. 106; Tromholt 1890, p. 20, p. 111; Tromholt 1892, p. 29, p. 133) is an illustration of two cognitive biases: "one puzzle—one solution" and "solution must be symmetric".

Two authors repeated the puzzle without adding fundamentally different, asymmetric solutions. One of them copied Tromholt (Botermans 2006, p. 12, p. 184). The other (Anonymous 2017, p. 61, p. 117) made unimportant changes: initial matchstick configuration and solution are rotated to left.

Formulation, initial matchstick configuration, and single solution to the first Tromholt's puzzle are given in Fig. 6.7.

Braun's strategy helps to find additional, fundamentally different (asymmetric) solutions. As the initial configuration is formed from 38 matchsticks, after removing 18 matchsticks, 20 matchsticks are left for forming 3 unequal square.

In the first class of fundamentally different asymmetric solutions, there are: one 3×3 square (12 matchsticks used) and two 1×1 squares (additional 8 matchsticks used). Two examples from such class of solutions are presented in Fig. 6.8.

Both of two solutions have as visually different solutions their mirror images.

In the second class of solutions there are:

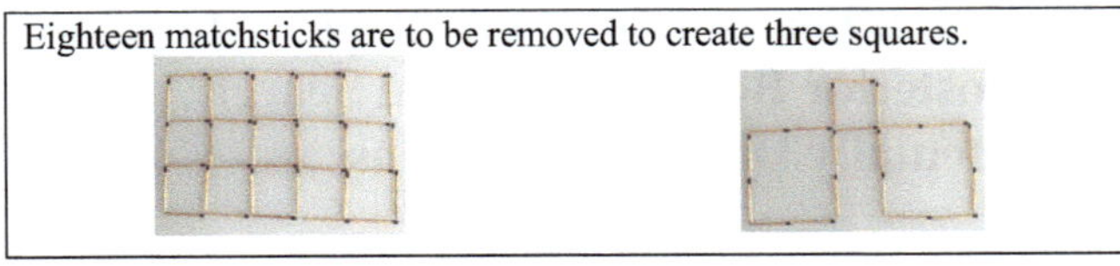

Fig. 6.7 Formulation, initial matchstick configuration, and symmetric solution of the first Tromholt's puzzle (*Source* Author)

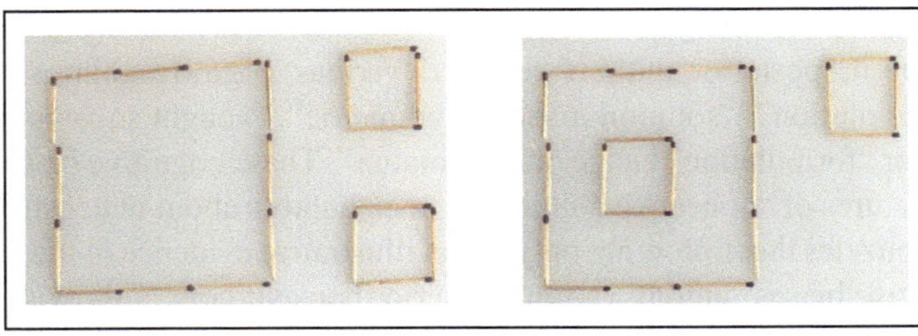

Fig. 6.8 Two examples of solution from the first class (*Source* Author)

One big square 3 × 3 (12 matchsticks used), one middle square 2 × 2 outside of the big square (additional 16 matchsticks used), and one small square 1 × 1 also outside the big square (additional 2 matchsticks used).

For these positions of squares 3 × 3 and 2 × 2, there are 9 different positions of small square.

In the third class of solutions there are two configurations:

One big square 3 × 3 (12 matchsticks used), one middle square 2 × 2 inside the big square (additional 4 matchsticks used), and one small square (1 × 1) outside the big square (additional 4 matchsticks used).

For these positions of squares 3 × 3 and 2 × 2, there are 3 different positions small square.

First configurations of these two classes of solutions are in Fig. 6.9.

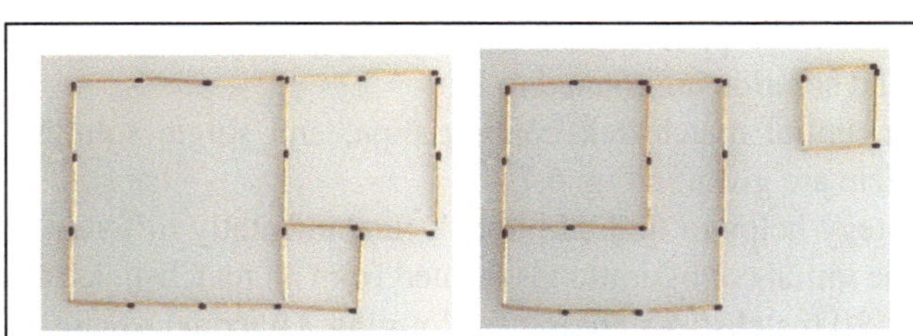

Fig. 6.9 First configurations of the second and third class of solutions (*Source* Author)

6.7 Remove 20 Matchsticks to Get 10 Squares

The second Tromholt's puzzle (Tromholt 1889, p. 25, p. 108; Tromholt 1890, p. 24, p. 112; Tromholt 1892, p. 33, p. 134) illustrates again the cognitive biases of "one puzzle—one solution" and "solution must be symmetric". It was repeated by two book authors (Wesel 1912, pp. 76–77; Paraquin 1970, p. 6, p. 27).

Formulation, initial matchstick configuration and solution to the second Tromholt's puzzle are presented in Fig. 6.10.

Braun's strategy is useful to find additional, fundamentally different asymmetric solutions. In initial matchstick configuration there are 60 matchsticks. After removing 20 matchsticks, for forming 10 squares one is left with 40 matchsticks.

1. The first configuration possibility to find fundamentally different solutions is: one 4 × 4 square (16 matchsticks used), four 2 × 2 squares (additional 8 matchstick used), and five 1 × 1 squares (additional 16 matchsticks used). One solution from the first configuration possibility is presented in Fig. 6.11.

This solution has its rotation variants.

2. The second configuration possibility to find fundamentally different solutions is: One 4 × 4 square (16 matchstick used), two 2 × 2 squares (additional 6 matchsticks

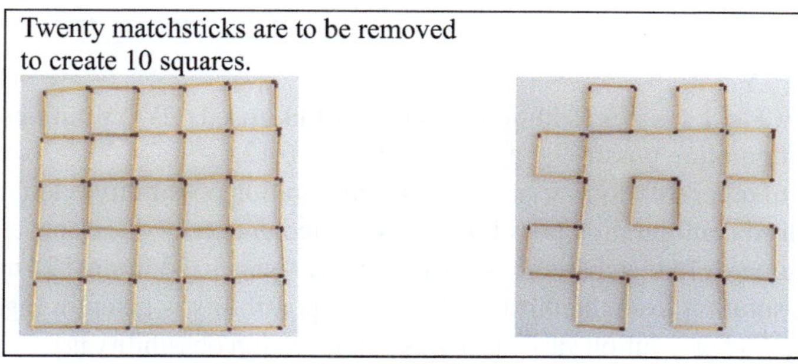

Fig. 6.10 Formulation, initial matchstick configuration, and of solution to the second Tromholt's puzzle (*Source* Author)

Fig. 6.11 One fundamentally different asymmetric solution from the first configuration possibility (*Source* Author)

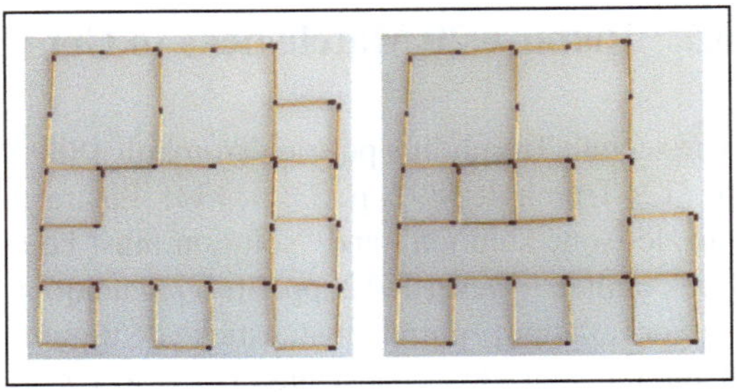

Fig. 6.12 Two fundamentally different asymmetric solutions from the second configuration possibility (*Source* Author)

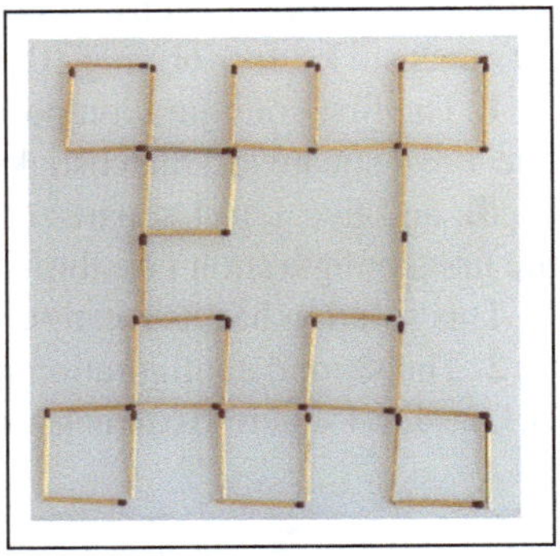

Fig. 6.13 One solution from the third configuration possibility (*Source* Author)

used), seven 1 × 1 squares (additional 18 matchsticks used). Two solutions from the second configuration possibility are shown in Fig. 6.12.

Finding of other solution variants in this configuration possibility is left to readers.

3. The third configuration possibility to find fundamentally different solutions is: One 3 × 3 square (16 matchsticks used), nine 1 × 1 squares (additional 24 matchstick used). One solution from the third configuration possibility is given in Fig. 6.13.

Finding of other solution variants in this configuration possibility is left to readers.

6.8 Remove 2 Matchsticks to Get 4 Squares

The third, very popular Tromholt's puzzle (Tromholt 1889, p. 35, p. 111; Tromholt 1890, p. 35, p. 115; Tromholt 1892, p. 44, p. 137) shows the presence of two cognitive biases: "one puzzle—one solution" and "sought squares must be equal".

Its formulation and solution were repeated by many posterior book authors (Czepa 1910, p. 16, p. 54; Lüders 1919, p. 33; Paraquin 1970, p. 3, p 26; Orleans and Orleans 1977, p. 13, p. 87; Clements 2006, p. 39).

Formulation, initial matchstick configuration and single solution to the third Tromholt's puzzle are given in Fig. 6.14.

Along with its rotation version, the given solution is the unique solution for unstated restriction that all four squares should be equal-sized.

Two matches are to be removed,
leaving 4 squares.

Fig. 6.14 Formulation, initial matchstick configuration, and solution to the third Tromholt's puzzle (*Source* Author)

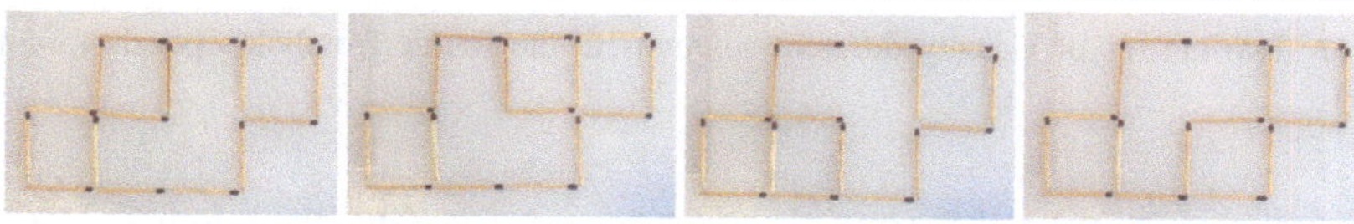

Fig. 6.15 Four additional solutions to the third Tromholt's puzzles (*Source* Author)

Nevertheless, those puzzle solvers, who escape two mentioned cognitive biases, would find that there are four additional, visually different solutions with one 2 × 2 square and three 1 × 1 squares. Two of these 1 × 1 squares are outside the 2 × 2 square and one 1 × 1 is inside it. Any one of these solutions is fundamentally different from single solution with equal squares found by Tromholt and posterior book authors. Four additional solutions with unequal squares are presented in Fig. 6.15.

6.9 Remove 6 Matchsticks to Get 4 Squares

The fourth Tromholt's puzzle (Tromholt 1889, p. 23, p. 108; Tromholt 1890, p. 23, p. 112; Tromholt 1892, p. 32, p. 134) again illustrates the presence of two cognitive biases: "one puzzle—one solution" and "sought squares are equal".

Its formulation and solution were repeated by many book authors (Czepa 1910, p. 14, p. 53; Wesel 1912, pp. 17–18; Blyth 1921, p. 41; Uhl 1947, p. 14, p. 40; Schuyt 1989, p. 63, p. 146; Obermair 1977, p. 21, p. 97).

Formulation, initial matchstick configuration, and single solution to third Tromholt's matchstick puzzle are shown in Fig. 6.16.

Six matchsticks are to be removed
to leave four squares.

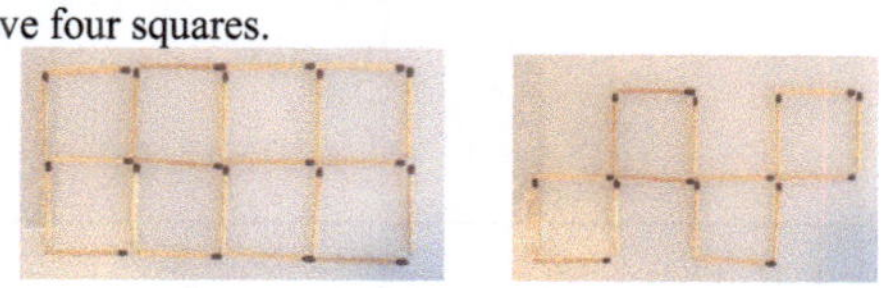

Fig. 6.16 Formulation, initial matchstick configuration, and solution to the third Tromholt's puzzle (*Source* Author)

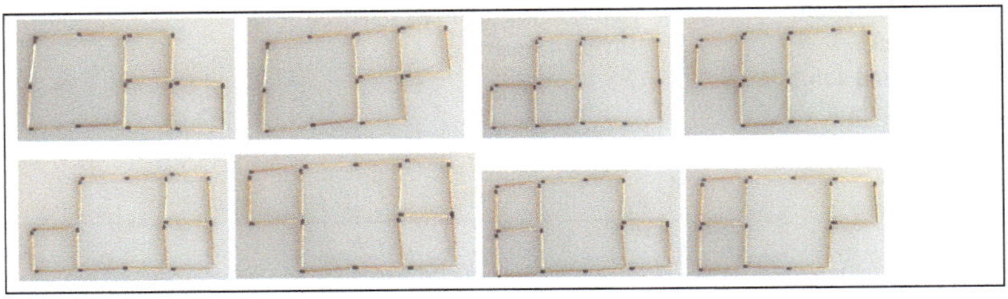

Fig. 6.17 Two classes of visually different solutions to the fourth Tromholt's puzzle (*Source* Author)

Just one author (Paraquin, 1970, p. 3, p. 26) recognized that Tromholt's solution is unique only if sought squares should be equal. Being so, this author reformulated the puzzle, adding that restriction.

For Tromholt's unrestricted puzzle formulation and its posterior repetitions, there are two classes of additional, visually different solutions. Both have four unequal-sized squares: one 2 × 2 square and three 1 × 1 squares. In one class, three 1 × 1 squares are together on the right or the left side of 2 × 2 square. In other class, two 1 × 1 squares and one 1 × 1 square are on different sides of 2 × 2 square. Two of these eight solutions (one from each class) are fundamentally different solutions. Two classes of solutions with four unequal squares are given in Fig. 6.17.

6.10 Move 6 Matchsticks to Get 4 Equal Squares

The fifth Tromholt's puzzle (Tromholt 1889, p. 53, p. 114; Tromholt 1890, p. 53, p. 117; Tromholt 1892, p. 63, p. 139) is example of cognitive bias "one puzzle—one solution". Its formulation and solution were repeated by Lüders (1919, p. 20, p. 21).

Formulation, initial matchstick configuration and single solution to fifth Tromholt's puzzle are given in Fig. 6.18.

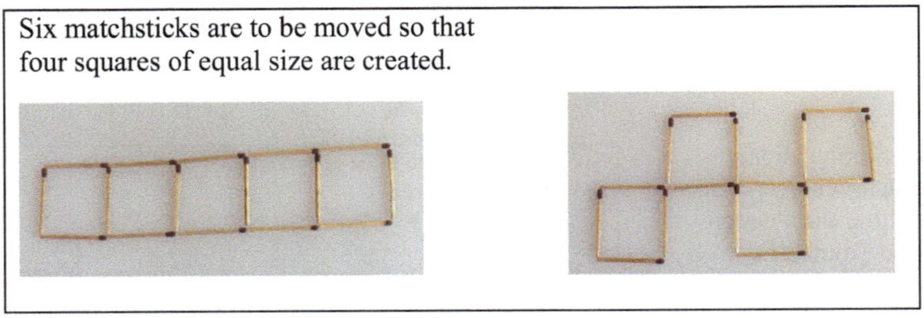

Fig. 6.18 Formulation, initial matchstick configuration, and solution to the fifth Tromholt's puzzle (*Source* Author)

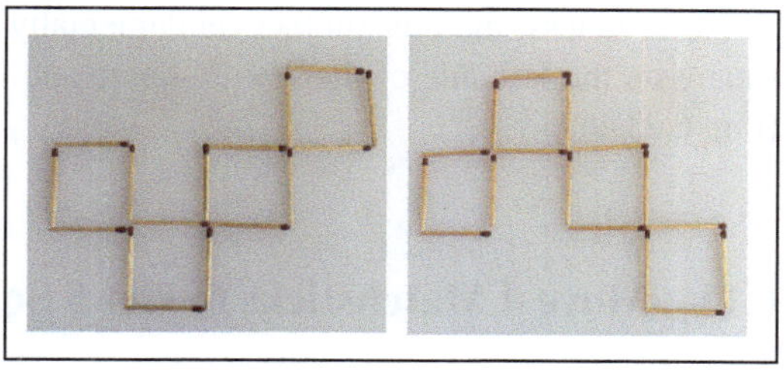

Fig. 6.19 Two Additional, fundamentally different solution to Tromholt's fifth puzzle (*Source* Author)

Due to cognitive bias "one puzzle—one solution", both Tromholt and Lüders missed two additional, fundamentally different solutions with 4 equal squares presented in Fig. 6.19.

Two posterior book authors (Paraquin 1970, p. 4; Anonymous 1996, p. 75, p 163) omitted in their formulations the restriction of equal-sized squares. Doing so, their reformulation gives to the puzzle more solutions, in one class are solutions with equal squares and in other class are solutions with unequal squares.

The first author revealed the cognitive bias "formulation change doesn't matter" and published the solution with four equal square which was mirror reflection of Tromholt's solution (Paraquin 1970, p. 26).

The second author avoided cognitive bias "formulation change doesn't matter", but not cognitive bias "one puzzle—one solution". In the book only one solution with four unequal squares was published (Anonymous 1996, p. 163). The published solution is given in Fig. 6.20.

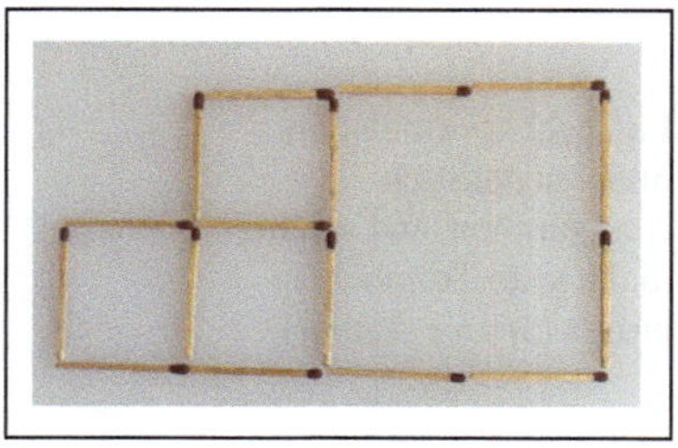

Fig. 6.20 The solution with three 1 × 1 squares on the left side of the 2 × 2 square (*Source* Author)

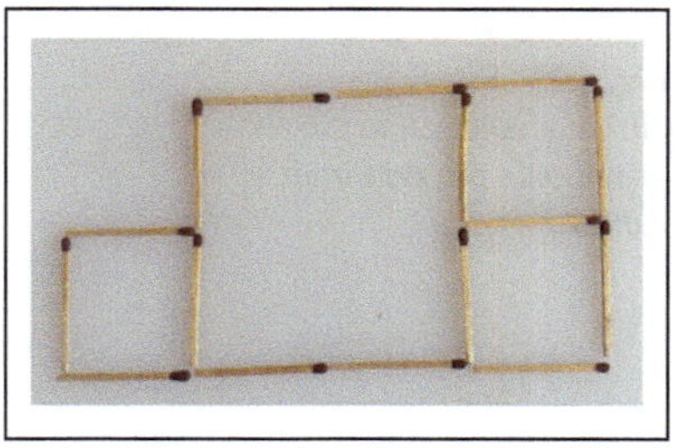

Fig. 6.21 Fundamentally different solution with one 1 × 1 squares on the left side of the 2 × 2 square and two 1 × 1 square on its right side (*Source* Author)

Nevertheless, there is another fundamentally different solution, with one 1×1 square on the left side of the 2×2 square and two 1×1 squares on its right side (Fig. 6.21).

6.11 Move 4 Matchsticks to Get 5 Squares

The sixth Tromholt's puzzle (Tromholt 1889, p. 34, p. 111; Tromholt 1890, p. 33, p. 114; Tromholt 1892, p. 42, p. 136) is another example of the most frequent cognitive bias "one puzzle—one solution". The puzzle and Tromholt's solution were repeated by various posterior book authors. (Lüders 1919, p. 34; Schuyt 1989, p. 52, p. 144; Botermans 2006, p. 112–113, p. 209).

Formulation, initial matchstick configuration and single solution to the sixth Tromholt's puzzle are shown in Fig. 6.22.

Neither Tromholt nor Lüders, Schuyt, and Botermans noticed that an additional, easy-to-find fundamentally different solution exists. This solution is given in Fig. 6.23.

Only one posterior book author, Alois Czepa, found a solution that is fundamentally different from Tromholt's solution (Czepa 1910, p. 30, p. 58). It is given in Fig. 6.24.

Although Czepa found solution, fundamentally different from the solution of Tromholt and other book authors, he also was unable to avoid cognitive bias "one puzzle—one solution". Otherwise, Czepa would have easily noted two different ways to form the five new equal squares from his solution (Fig. 6.25).

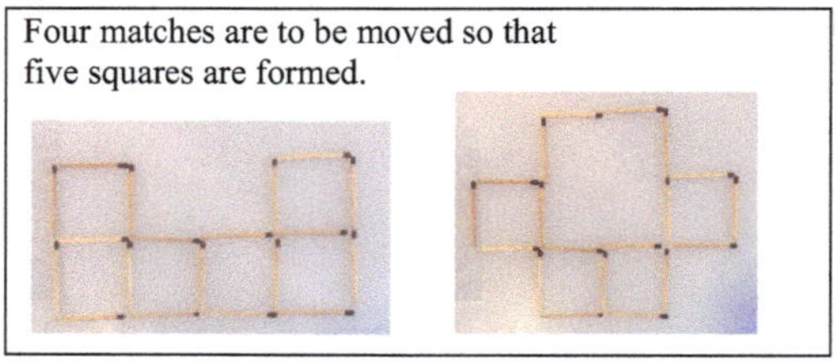

Fig. 6.22 Formulation, initial matchstick configuration, and solution to the sixth Tromholt's puzzle (*Source* Author)

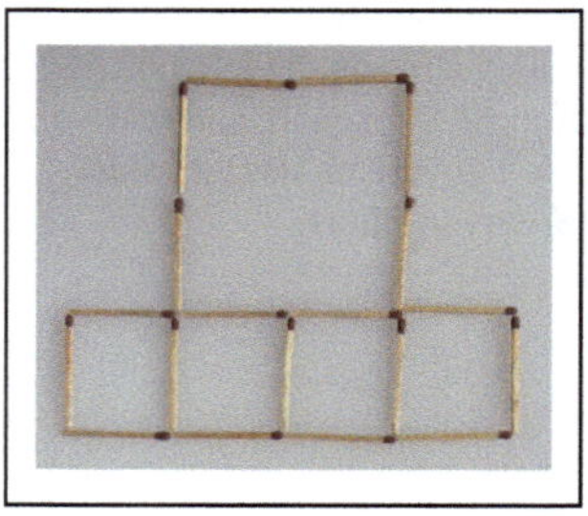

Fig. 6.23 Fundamentally different solution to the sixth Tromholt's puzzle (*Source* Author)

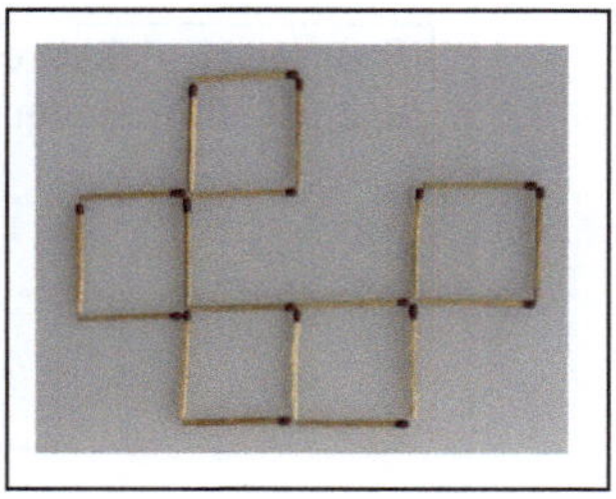

Fig. 6.24 Czepa's fundamentally different solution (*Source* Author)

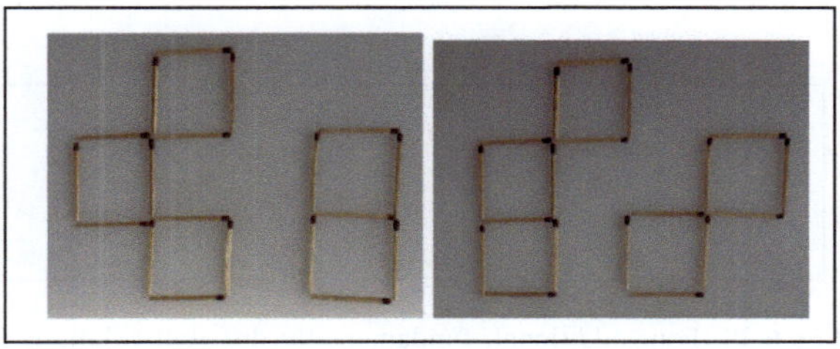

Fig. 6.25 Two easy-to-find solutions which might be derived from Czepa's solution (*Source* Author)

Ten solutions with 5 equal squares can be seen in my book on matchstick puzzle published in Croatia (Slisko 2024, pp. 215–216).

6.12 Remove 4 Matchsticks to Get 3 Squares

The seventh Tromholt's puzzle (Tromholt 1892, p. 112, p. 148) is one more example of cognitive bias "one puzzle—one solution".

This puzzle was previously published by Lois Mittenzwey. His formulation, initial matchstick configuration, and solution (Mittenzwey 1880, p. 37, p. 90) are given in Fig. 6.26.

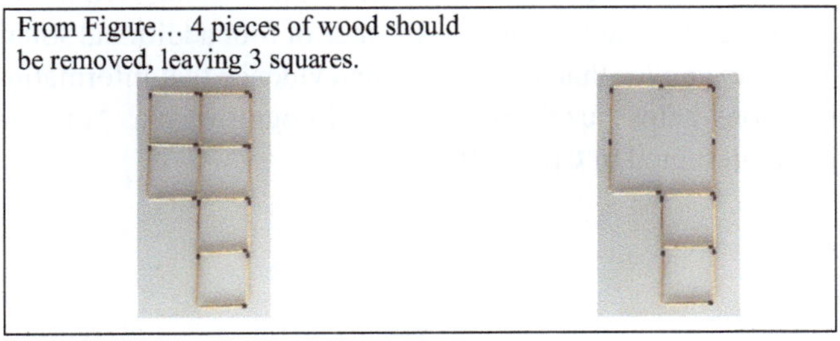

Fig. 6.26 Formulation, initial matchstick formulation, and solution of Mittenzwey's puzzle. (*Source* Author)

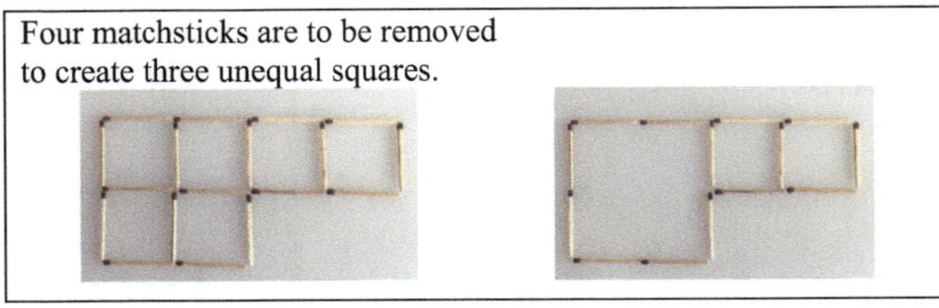

Fig. 6.27 Formulation, initial matchstick formulation, and solution of the seventh Tromholt's puzzle. (*Source* Author)

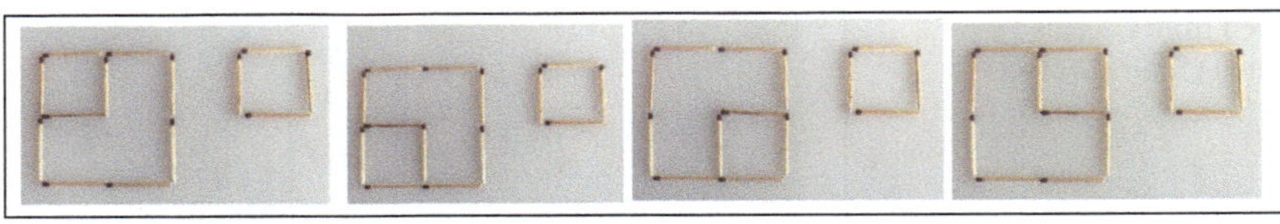

Fig. 6.28 Four additional, fundamentally different solutions for the seventh Tromholt's puzzle (*Source* Author)

Tromholt, 12 years later, without giving credit to Mittenzwey, repeated the puzzle, rotating the initial matchstick configuration to left. Tromholt's formulation, initial matchstick formulation, and solution are presented in Fig. 6.27.

It is important to observe that Tromholt was more specific about squares' feature ("unequal squares").

Czepa (1910, p. 47, p. 63) and Paraquing (1970, p. 3, p. 26) repeated Tromholt's puzzle and solution. By publishing only one solution to this puzzle, all authors again didn't escape cognitive bias "one puzzle—one solution".

Creative persons, who have inclination to think about the possibilities of alternative solutions to a puzzle, would very likely find four fundamentally different solutions, given in Fig. 6.28.

In a small-scale study in Serbia, carried out by Ivana Krulj, three out of twenty-one 14-year students were able to find them, after seeing Tromholt's solution (Fig. 6.27) and being informed about the existence of four additional solutions. Other experiment-based examples that are convincing evidence that information about the number of solutions helps puzzle solvers avoid cognitive bias "one puzzle—one solution" will be presented in Chap. 10.

6.13 Make 2 Squares and 4 Right Triangles from 8 Matchsticks

The eight Tromholt's puzzle shows that cognitive bias of "uncritical thinking" exists, too! That puzzle was likely inspired by a geometric puzzle published by Lois Mittenzwey (Mittenzwey 1880, p. 35, p. 88). Formulation and solution of Mittenzwey's puzzle are given in Fig. 6.29.

It is easy to detect that Mittenzwey's solution drawing is incorrect (cognitive bias "uncritical thinking") because the side length of the inner square isn't 1 cm but is 2 cm (half of the triangle side with length of 4 cm)!

Without noticing that error, Tromholt transformed Mittenzwey's geometric puzzle into a matchstick puzzle (Tromholt 1889, p. 65, p. 114). Formulation and solution of that Tromholt's puzzle are given in Fig. 6.30.

Tromholt's solution is incorrect because implies that the length of one cathetus is equal to the length of the hypotenusa!

This puzzle wasn't included in the fourth edition (Tromholt 1890). It is unknown whether Tromholt himself discovered that the solution was incorrect or he was informed about the error by a critical-minded or a practical reader. Nevertheless, "the ghost was out of bottle" and ironically this Tromholt's puzzle with erroneous solution became his most popular matchstick puzzle!

Here come only a few examples of repeated Tromholt's puzzle:

> Arrange 8 matches so as to form two squares and four triangles. (Anonymous 1917, pp. 42–43).
>
> Form two squares and four triangles with eight matches. (Bowers 1934, p. 84).
>
> Using eight matches, form two squares and four triangles. (Masson and Mitchel 1935, p. 180)
>
> Form 2 squares and 4 triangles with 8 matches (Green 1955, p. 104)
>
> Using eight matches, form two squares and four triangles. (Krau 1960, p. 69)

Make the following figures: 1) a square with sides measuring 1 cm and 2) four right-angled triangles with one side measuring 3 cm, the second 4 cm and the third 5 cm and put these 5 figures together to form a square!

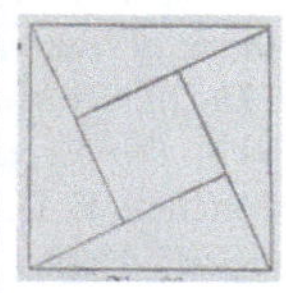

Fig. 6.29 Formulation and solution of Mittenzwey's puzzle (*Source* Mittenzwey 1880)

Fig. 6.30 Formulation and solution of Tromholt's eight puzzle (*Source* Tromholt 1889)

How can you make a square consisting of 1 square and 4 triangles using 8 matchsticks?

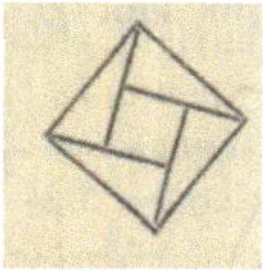

Arrange eight matches to form two squares and four triangles. (Brooke 1973, p. 7, p. 42)

Using 8 matchsticks, make 4 right triangles the same size and 2 squares of different size. (Obermair 1977, p. 33, p. 114)

Make two squares and four triangles from eight matches, without breaking or bending them (Fixx 1982, p. 28, p. 108)

References

Anonymous (1917). *Boys' Life. The Boy Scouts' Magazine, 7*(1), 42–43.

Anonymous (1996). *Streichholz-Spiele.* Tosa Verlag.

Anonymus (2014). *Das fabelhafte Büchlein der besten Streichholzspiele.* Regionalia Verlag.

Anonymous (2017). *Matchstick Puzzles. Test your brain power with these challenging puzzles.* Arcturus Publications.

Blyth, W. (1921). M*atch-stick magic.* C. Arthur Pearson.

Botermans, J. (2006). *Matchstick Puzzles.* Sterling Publishing Company.

Bowers, E. (1934). *Recreation for girls and women.* A.S. Barnes and Company.

Brooke, M. (1973). *Tricks, games, and puzzles with matches.* Dover Publications.

Clarke, R. J. (2017). *Matchstick Puzzles.* CreateSpace.

Clements, S. (2006). *Thinking skills. Daily brainteasers for ages 9–11.* Scholastic.

Colson, R. (2012). *Perplexing puzzles, cryptic challenges, remarkable riddles. Navigate a labyrinth of over 300 puzzles & riddles.* Parragon.

Czepa, A. (1910). *Streichholz-Spielereien: Kunststücke und Scherze nebst Auflösungen.* Otto Maier Verlag.

Fixx, J. F. (1982). *Games for the super-intelligent.* Warner Books.

Green, A. V. (1955). *Simple Tricks.* Doubleday & Company.

Grund-Thorpe, H. (2006). *Die schönsten Streichholzspiele.* Weltbild Verlag.

Guthrie, J. & Preston, T. (2017). *Boost Your Brainpower. Over 400 Puzzles, Games, and Brain Teasers to Boost Your Mind.* Charterwell Books.

Hansell, S. (1981). *109 knifflige Streichholztricks.* Otto Maier Verlag, Ravensburg.

Huang, A., Lan, D. & Huang, B. (2020*). The Art of Problems Solving. Good foundation of logic and programming. C level - Book 1.* 99 Learning LLC.

Johnson-Laird, P. N. (2005). The shape of problems, in Girotto, V. & Johnson-Laird, P. H. (Editors) (2005). *The shape of reason. Essays in honor of Paolo Legrenzi.* Psychology Press.

Krau, R. (1960). *The family book of games.* McGraw-Hill Book Company.

Lüders, C. E. (1919). *Der Streichholzkünstler in der Westentasche.* Verlag G. Danner.

Mason, B. S. & Mitchell, E. D. (1935). *Social games for recreation.* A. S. Barnes and Company.

Mittenzwey, L. (1880). *Mathematische Kurzweil.* Julius Klinkhardt.

Obermair, G. (1977). *Matchstick puzzles, tricks and games.* Sterling Publishing Company,

Orleans, S. & Orleans, J. (1977). *Pencil puzzles 5.* Grosset & Dunlap.

Orleans, S. & Orleans, J. (1994). *The bumper puzzle book.* Chancellor Press.

Paraquin, K. H. (1970). Spiel und Spass mit Hölzchen und Köpfchen, Otto Maier Verlag.

Picon, D. (2002). *Allumettes.* Mango.

Schuyt, M. (1989). *Phantastische Zündholzspiele.* DuMont Buchverlag.

Seymoure, D. (1991). *Mathematics. Exploring your world. Thinking critically.* Silver Burdett & Ginn.

Sliško, J. (2024). *Zagonetke sa šibicama. Zadaci, upute i rješenja za roditelje i učitelje.* Element.

Tromholt, S. (1889). *Streichholzspiele. Denksport und Kurzweil.* First edition. Otto Spamer Verlag.

Tromholt, S. (1890). Streichholzspiele. *Denksport und Kurzweil.* Fourth edition. Otto Spamer Verlag.

Tromholt, S. (1892). *Streichholzspiele. Denksport und Kurzweil.* Fifth edition. Otto Spamer Verlag.
Van Delft, P. & Botermans, J. (1978). *Creative puzzles of the world.* Cassel.
Uhl, C. (Editor) (1947). *Der Kindergarten. Heft 8. Streichholzspiele.* Verlag Werden und Wirken.
Wesel, M. (1912). *Streichholzscherze in der Westentasche.* Verlag G. Danner.
Williams, B. (2013). *Gymnastics for the brain. Matchstick puzzles.* Salim Bouzekouk

Chapter 7
Restricted and Unrestricted Formulations of Matchstick Puzzles

Abstract The aim of this chapter is to show the evidence that features "one puzzle—one solution", "careless formulation change", and "incomplete solution set" appear also in popular matchstick puzzles that were published by authors who came after Tromholt. It means that these features, caused by different cognitive biases, aren't accidental behavior of a single author but rather a collective behavior in the community of authors write books on matchstick puzzle. In the first two puzzles, special attention is paid to unfortunate phenomenon: authors, who carelessly reformulate puzzles, eliminating the restrictions about the size-equality of remaining triangles or squares, publish the solution to puzzles with restrictive formulations. Namely, unrestrictive formulations of puzzle open possibilities for multiple solutions, either visually or fundamentally different. For the third, multiple-solution puzzle either in restricted or unrestricted formulation, all book authors published only different version of a single solution.

7.1 Remove 4 Matchsticks to Get 4 Equal Triangles

Henry Ernest Dudeney, famous British puzzle-maker, published in the Year 1925 his most popular matchstick puzzle. Its formulation, initial matchstick configuration, and solution (Dudeney 1925a, p. 634; Dudeney 1925b, pp. 107–108) are given in Fig. 7.1.

This puzzle was later repeated in Dudeney's books (Dudeney 1932, p. 116, p.185; Dudeney 1967, p. 206, p. 406) and in books written by other authors (Fujimura 1978, p. 9, p. 109; Larson et al. 1998, p. 168; Colson 2012, p. 59, p. 191; Anonymous 2015, p. 43, p. 111).

Its destiny took an additional and different path in the Year 1961, when Evans made two changes in Dudeney' puzzle (Evans 1961, p. 266, p. 271). Formulation, initial matchstick configuration, and solution to Evans' puzzle are given in Fig. 7.2.

The first change was the rotation to the left of Dudeney's initial matchstick configuration. The second and more important change was to eliminate the restriction

J. Slisko, *Fostering Cognitive Mathematics Skills with Matchstick Puzzles*, Mathematics in Mind, https://doi.org/10.1007/978-3-032-14561-1_7

A little match puzzle Here is a little puzzle for the more juvenile reader. Place sixteen matches, as shown, to form eight equilateral triangles. Now take away four matches to leave only four equal triangles. No superfluous matches or loose ends to be left.

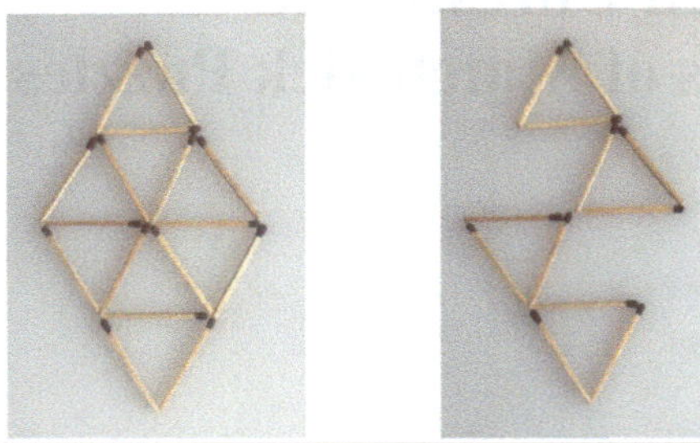

Fig. 7.1 Formulation, initial matchstick configuration, and solution of Dudeney's puzzle (*Source* Author)

Lay down sixteen matches to form ten triangles. Remove four matches and leave only four triangles.

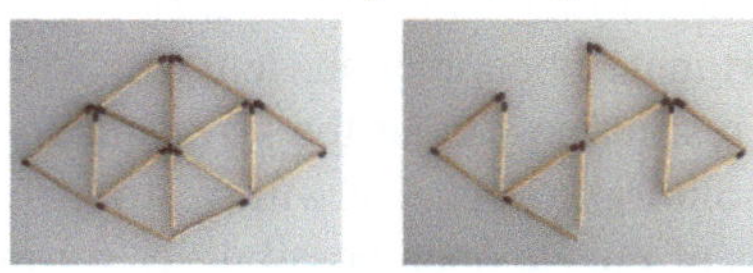

Fig. 7.2 Formulation, initial matchstick configuration, and solution to Evans' puzzle (*Source* Author)

that remaining triangles must be of equal size. Nevertheless, Evans' published only solution with equal-sized triangles!

Being so, Evans didn't escape three cognitive biases, that are frequently present in the world of matchstick puzzles. The first cognitive bias is "one puzzle—one solution". The second cognitive bias is "formulation change doesn't matter". The third cognitive bias is "remaining triangles are equal", although elimination of equal-size restriction opens the possibility for many solutions with unequal triangles.

After Evans published his matchstick configuration, it became very popular and was accepted by many book authors, although they behaved in three different ways.

A few of them in their formulations kept the restriction that the remaining triangles should be equal or congruent (Bolt 1985, p. 10, p. 114; Battaglia 2002, p. 6; Rapoport and Yoder 2017, pp. 94–95, p. 137). Townsend (1990, p. 20, p. 80) made a strange erroneous combination. He had Dudeney's formulation and initial stick configuration, while the solution was within the configuration of Evans!

Bigger group of book authors followed Evans, omitting in their formulations the restriction of equal-sized triangles and publishing only Evans' solution with four equal triangles (Brooke 1973, p. 26, p. 56; Brandreth 1977, p. 61, p. 67; Brandreth 1982, p. 127, p. 183; Eldin 1982, p. 8, p. 96; Slocum 1996, p. 21, p. 41) or its mirror image (Vogt 1994, p. 4, p. 70).

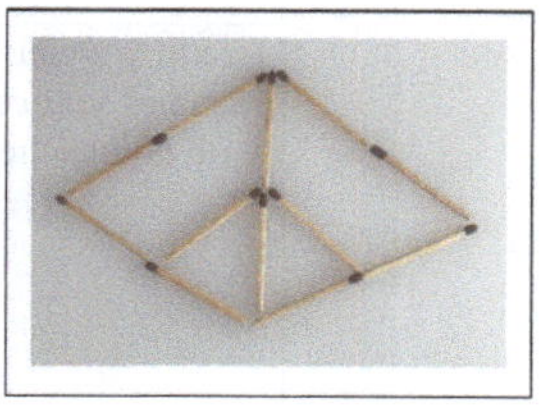

Fig. 7.3 The only solution with unequal triangles which was found by all book authors who eliminated equal-size restriction (*Source* Author)

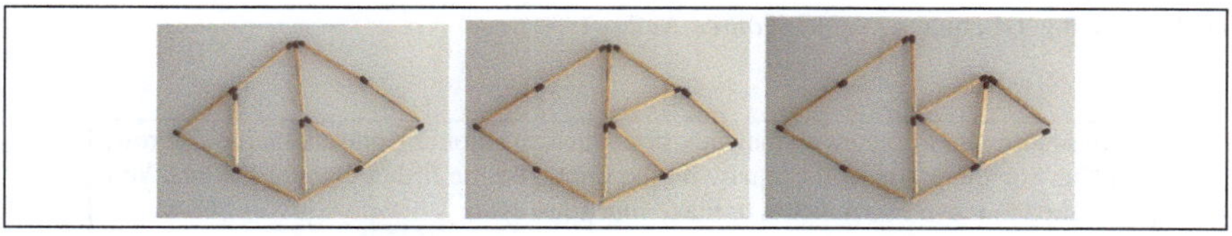

Fig. 7.4 Three fundamentally different solutions with unequal triangles which can be constructed from the only one published by all book authors (*Source* Author)

Catherine S. Ross opted for a new approach. She eliminated equal-size restriction for Dudeney's configuration and published only one solution which is mirror image of Dudeney's solution (Ross 2014, p. 77, p. 181).

Only in one book, that repeated the Evans' puzzle, the single published solution is with unequal triangles (Hardy et al. 2001, p. 32, p. 203). It is given in Fig. 7.3.

After having found this symmetrical solution, persons with visual intelligence, who don't believe that matchstick puzzles must have only one solution, would likely find more fundamentally different solutions. Three of those possible solutions are presented in Fig. 7.4.

Twelve different solutions to the Evans' puzzle found by Hrvoje, a high-school student from Croatia, will be presented in Chap. 10.

7.2 Move 2 Matchsticks to Get 4 Equal Squares

The first publication of this very popular puzzle, in which five squares should be changed into four squares by moving two matchsticks, appeared in the Year 1937 in the journal "*The Iron Age*". Its formulation, initial matchstick configuration, and solution (Anonymous 1937a, p. 137; Anonymous 1937b, p. 119) are given in Fig. 7.5.

In the Year 1951, Kohl and Young omitted the restriction that the remaining square should be equal-sized. Nevertheless, they kept the single solution with the four equal squares. Their formulation, initial configuration, and solution (Kohl and Young 1951, p. 80) are presented in Fig. 7.6.

The treatments of puzzle by posterior book and journal authors can be divided into three different groups.

Credit yourself with uncommon acuteness if by replacing two of the matches you can change the five squares into four of the same size, with no matches left over.

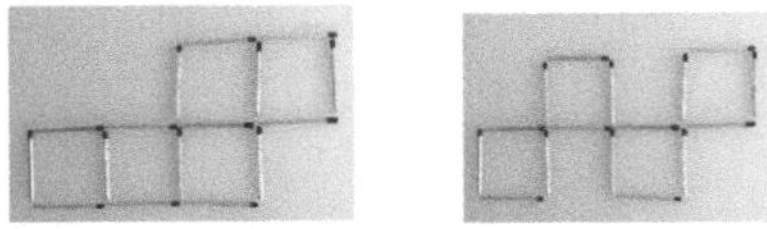

Fig. 7.5 Formulation, initial matchstick configuration, and solution of the puzzle "Move 2 matchsticks to get 4 equal squares" (*Source* Author)

Five squares are formed with matches, three squares adjoining one another in one row, another square on top of the square on the right, and the fifth adjoining the last. You are asked to move two matches to make four squares.

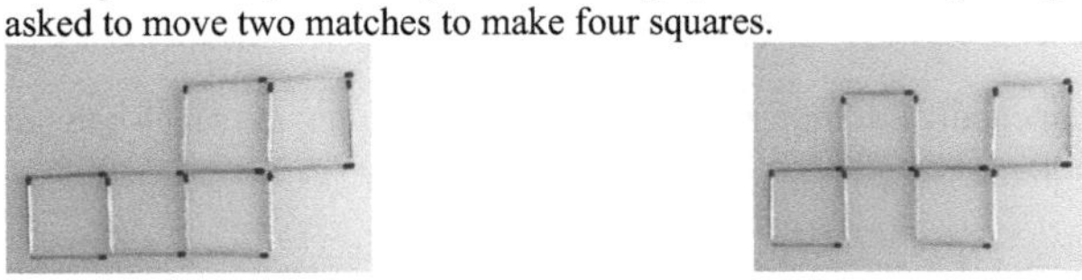

Fig. 7.6 Formulation, initial matchstick configuration, and solution of the puzzle "Move 2 matchsticks to get 4 squares" without restriction of their equality (*Source* Author)

To the first group belong the authors who included in their formulations that the remaining four squares should be equal (Harbin 1963, p. 92, p. 109; Holt 1981, p. 60, p. 93; Holt 1983, p. 18, p. 81; Davidson 2000, p.106; Picon 2002, p. 20, p. 116; Grund-Thorpe 2006, p. 22, p. 87; Cutler 2015, p. 152).

In the second group are the authors who, following the approach of Kohl and Young, eliminated the restriction that four remaining squares should be equal. Nevertheless, they published only one solution with four equal squares (Rigney 1960, p. 2; Brooke 1973, p. 18, p. 50; Eldin 1982, p. 9, p. 96; Ward 1986, p. 11, p. 30; Logan 1991, p. 214; Bolt 1992, p. 1, p. 73; Klein 1999, p. 44, p. 63; Allen et al. 2005, p. 115, p. 364; Harrison and Hobbs 2010, p. 80, p. 176; Overholt and Kincheloe 2010, p. 370, p. 371; Bragdon 2011, p. 169, p. 191; Anonymous 2014, p. 38, p. 102). Due to cognitive bias "formulation change doesn't matter", they didn't notice that in such, unrestricted formulation there are additional solutions with unequal squares.

In the third group are the authors who noticed that unrestricted formulation implies that there can be one or two solutions with four unequal squares. James Fuller Fixx was the first author to publish one solution with unequal squares, announcing squares inequality even in the puzzle formulation. His formulation, initial matchstick configuration, and solution (Fixx 1972, p. 12, p. 71) are in Fig. 7.7.

Some authors repeated the formulation and solution of Fixx (Anonymous 1986, p. 3, p. 15; Ellis and Young 2012, p. 226, p. 234), but others published only one solution with unequal squares, without information about inequality of squares in puzzle formulation (Robin 1976, p. 38, p. 118; Haggerty 1995, p. 34, p. 37).

Move two matches and make four squares, one of them larger than the other three.

Fig. 7.7 Formulation, initial matchstick configuration, and solution to Fixx's puzzle (*Source* Author)

Pasen and Roll published two fundamentally different solutions with unequal squares. Their formulation, initial matchstick configuration, and two solutions (Pasen and Roll 1975, p. 142, p. 279) are given in Fig. 7.8.

Gilbert Obermair was the first author who shown that two different puzzle formulations lead to two different classes of solutions (one with four equal squares and other with four unequal squares). Formulations, initial matchstick configuration, and solutions of Obermair's two-puzzle approach (Obermair 1977, pp. 20–21, p. 120, p. 138) are given in Fig. 7.9.

Using a different initial matchstick configuration, Jack Botermans repeated Obermair's two-puzzle approach (Botermans 2006, pp. 116–117, p. 210).

The other way for getting two different classes of solutions (with four equal and unequal squares) was by using a hint (Anonymous 1980, pp. 6–7, p. 32). It is presented in Fig. 7.10.

Move two matches and make four squares.

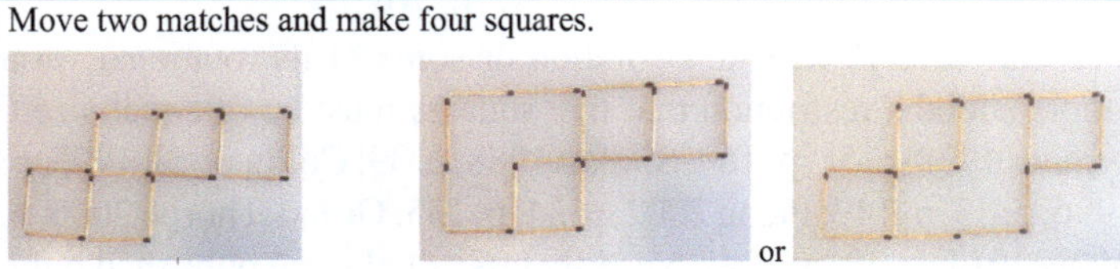

Fig. 7.8 Formulation, initial matchstick configuration, and solutions to the puzzle of Pasen and Roll (*Source* Author)

a) Move 2 matchsticks to make 4 equal-sized squares.
b) Move 2 matchsticks to make 4 squares – 3 small and one large.

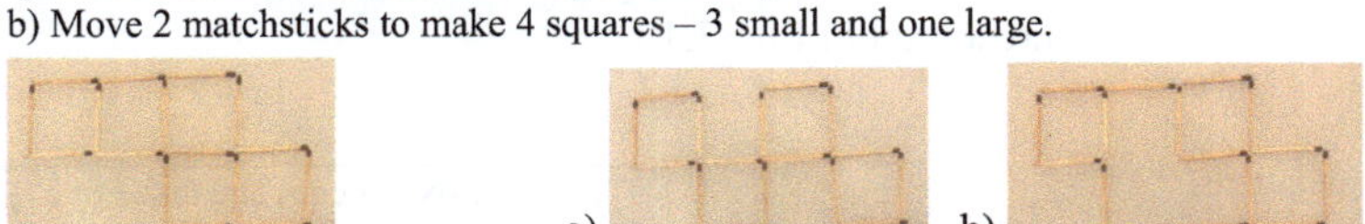

Fig. 7.9 Formulations, initial matchstick configuration, and solutions in two-puzzle approach of Obermair (*Source* Author)

Use 16 matches to make five squares, as shown (below). Then move two matches to new positions to form exactly four squares. **Hint** – There may be more than one solution.

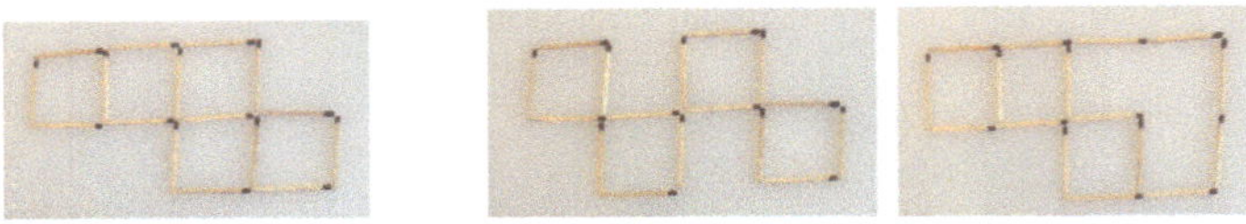

Fig. 7.10 Formulations, initial matchstick configuration, and solutions of the puzzle with hint (*Source* Author)

One author (Anonymous 1996, p. 15, p.137) even considers that the readers don't need a hint to find two different solutions (one with four equal and one with unequal squares).

7.3 Remove 6 Matchsticks to Get 5 Squares

This puzzle was published by Maxey Brooke. Its unrestricted formulation, initial matchstick configuration, and solution are in Fig. 7.11.

It is easy to notice that Brooke didn't escape the cognitive bias "all squares must be of equal size". Many book authors were misled by the same cognitive bias (Brown 1978, p. 39, p. 157, Brandreth 1981, Eldin 1982, pp. 11–12, p. 100; Brandreth 1984, p. 59, p. 231; Townsend 1994, p. 24, p. 109; Botermans 2006, pp. 8–9, p. 182; Bodycomb 2007, p. 280, p. 539; Roffy 2014, p. 67, p. 70).

In order to repeat "safely" Brooke's solution or some of its rotational variants, other book authors added a restriction that five squares must be of equal size (Van Delft and Botermans 1978, p. 51, p. 189; Straker 1988, p. 39; Collins et al. 1995, p. 83; Anonymous 1996, p. 43, p. 149; Picon 2002, p. 34, p. 135; Grund-Thorpe 2006, p. 21, p. 86; Anonymous 2014, p. 42, p. 104). Nevertheless, all of them published only one solution (cognitive bias "one puzzle—one solution"), apparently being uninterested in finding additional, fundamentally different solutions with 5 equal squares. Three of them are given in Fig. 7.12.

In similar way, the authors, whose unrestricted formulations permit many solutions with five unequal squares, were unable to find single one with such a configuration. Three of such solutions are presented in Fig. 7.13.

Fig. 7.11 Formulations, initial matchstick configuration, and solutions of Brooke's puzzle (*Source* Author)

Remove six matches and leave five squares.

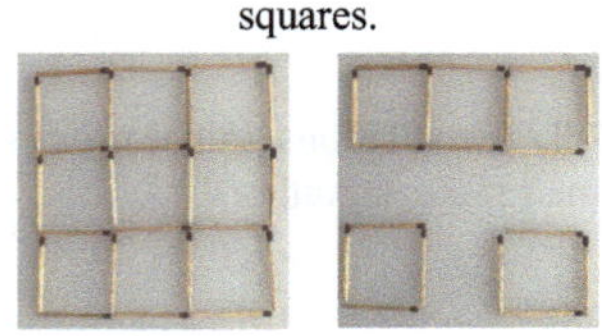

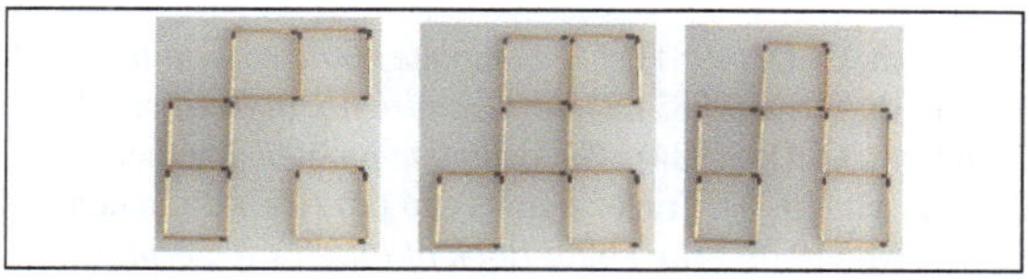

Fig. 7.12 Three possible fundamentally different solutions to Brooke's puzzle with five equal squares (*Source* Author)

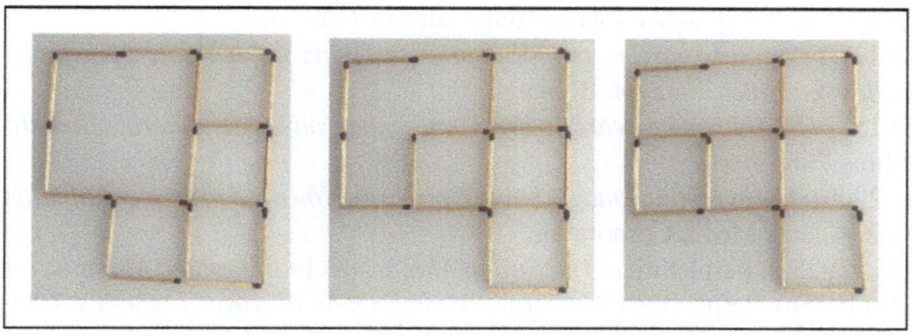

Fig. 7.13 Three possible, fundamentally different solutions to Brooke's puzzle with five unequal squares (*Source* Author)

It seems that in the case of Brooke's puzzle, either in its restricted or unrestricted formulation, all book authors didn't have enough motivational inspiration and/or creative skills for looking and finding more than one solution!

References

Bragdon, A. D. (Editor) (2011). *Mind Stretchers*. Jade Edition. Reader's Digest.

Allen, R., Ballheimer, D., Harrod, J. & Paines, J. (2005). *The biggest book of games for one ever! Over 500 games of luck, skill, and patience for players of a solitary disposition.* Barron's.

Anonymous (A.H.D.) (1937a). Problem. *The Iron Age, 140*(7), 137.

Anonymous (A.H.D.) (1937b). Solution, *The Iron Age, 140*(9), 119.

Anonymous (1980). Something to think about. *VCU magazine, 9*(2), 6–7, 32.

Anonymous (1986). Tech teasers. *Illinois Technograph, 101* (4), 3, 15.

Anonymous (1996). *Streichholz-Spiele.* Tosa Verlag.

Anonymous (2015). *Matchstick puzzles. Test your brain power with these challenging puzzles.* Arcturus.

Anonymus (1996). *Streichholz-Spiele.* Tosa Verlag.

Anonymus (2014). *Das fabelhafte Büchlein der besten Streichholzspiele.* Regionalia Verlag.

Battaglia, P. (2002). *So you think you're smart. 150 fun and challenging brain teasers.* International Puzzle Feature.

Bodycombe, D. J. (2007). *The riddles of the sphinx: and the puzzles, word games, brainteasers, conundrums, quizzes, mysteries, codes and ciphers that have baffled, entertained and confused the world over the last 100 years.* Penguin, p. 280, p. 539.

Bolt, B. (1985). *More mathematical activities. A resource book for teachers.* Cambridge University Press.

Bolt, B. (1992). *Mathematical cavalcade.* Cambridge University Press.

Botermans, J. (2006). *Matchstick Puzzles,* Sterling Publishing Company.

Brandreth, G. D. (1977). *Home entertainment for all the family.* The Stephen Greene Press.

Brandreth, G. D. (1981). *The complete home entertainer*. Hale.
Brandreth, G. D. (1982). *Gyles Brandreth's Book of puzzles and brainteasers*. Octopus.
Brandreth, G. D. (1984). *The complete puzzler*. Panther.
Brooke, M. (1973). Tricks, games and puzzles with matches. Dover Publications.
Brown, M. (1978). *The Great Book of Puzzles & Perplexities*. Lyceum Books, p. 39, p. 157.
Collins, W., Dritsas, L., Frey-Mason, P., Howard, A. C., McClain, K., Moore-Harris, B., Ott, J., Pelfrey, R. S., Price, J., Smith, B. & Wilson, P. S. (1995). *Mathematics. Applications and connections*. Course 2. Teacher's wraparound edition. Glencoe.
Colson, R. (2012). *Perplexing puzzles, cryptic challenges, remarkable riddles. Navigate a labyrinth of 300 puzzles & riddles*. Paragon Books.
Cutler, T. (2015). *The pilot who wore a dress and other dastardly lateral thinking mysteries*. HarperCollins Publishers.
Davidson, G. (2000). *The everything magic book. Everything you need to amaze, baffle, and entertain your friends*. Adams Media Corporation.
Dudeney, H. E. (1925a). Perplexities. *The Strand Magazine. An illustrated Monthly, 69*(6), 634.
Dudeney, H. E. (1925b). Perplexities. *The Strand Magazine. An illustrated Monthly, 70*(1), 107–108.
Dudeney, H. E. (1932). *Puzzles and curious problems*. Thomas Nelson and Sons.
Dudeney, H. E. (1967). 536 *puzzles and curious problems*. Charles Scribner's Sons.
Eldin, P. (1982). Match play. Safe puzzles, games and tricks with matchsticks. Granada Publishing.
Eller, D. & Young, S. (Editors) (2012). *7 years younger. The revolutionary 7-week anti-aging plan*. Hearst Magazines.
Evans, J. R. (1961). *The junior week-end book*. Victor Gollancz Limited.
Fixx, J. F. (1972). *Games for the Superintelligent*. Doubleday & Company.
Fujimura, K. (1978). *The Tokyo puzzles*. Charles Scribner's Sons.
Grund-Thorpe, H. (2006). *Die schönsten Streichholzspiele*, Weltbild Verlag.
Haggerty, B. A. (1995). *Nurturing intelligences. A guide to multiple intelligences theory and teaching*. Addison - Wesley.
Harbin, R. (1963). *Party lines*. Oldbourne Book Company.
Hardy, G., Blum, R., Ryan, S., Sukach, J., Moreau, R., Smith, S. & Sloane, P. (2001). *Giant book of puzzles*. Sterling Publishing Company.
Harrison, J. & Hobbs, M. (2010). *Brain training. Complete visual program*. Dorling Kinderslay.
Holt, M. (1981). *Figure it out 3*. Dragon Books.
Holt, M. (1983). *Doctor Who quiz book of magic*. Methuen Children's Books.
Klein, L. (1999). *Mazes and Mysteries*. Scholastic.
Kohl, M. & Young, F. (1951). *Games from grownups. The key to successful home entertainment with unique, handy charts*. Cornerstone Library.
Larson, R., Boswell, L. & Stiff, L. (1998). *Geometry. An integrated approach*. McDougal Littrell.
Logan, F. A. (1991). *College learning. Ways & whys*. Kendall/Hunt.
Obermair, G. (1977). *Matchstick puzzles, tricks and games*. Sterling Publishing Company.
Overholt, J. & Kincheloe, L. (2010). *Math Wise! Over 100 hands-on activities that promote real math understanding. Grades K-8*. Second Edition. Jossey-Bass.
Pasen, R. M. & Roll, J. H (1975). *Student study guide to accompany Psychology. Explorations in behavior and experience by Mednick, Higgins, Kirschenbaum*. John Wiley and Sons.
Picon, D. (2002). *Allumettes*. Mango.
Rapoport, R. & Yoder, J. A. (2017). *Math lab for kids: fun, hands-on activities for learning with shapes, puzzles, and games*. Quarto Publishing Group.
Rigney, F. J. (1960). *Cub Scout magic*. Boy Scouts of America.
Robin, S. (1976). *Recyclopedia. Games, science equipment, and crafts from recycled materials*. Houghton Mifflin.
Roffy (2014). *Laugh on the loo*. Paragon.
Ross, C. S. (2014). *Shapes in math, science and nature. Squares, triangles and circles*. Kids Can Press.
Slocum, J. (1996). *The puzzle arcade for people who like lots of hints*. Klutz.

Straker, A. (1988). *Home maths. Ages 10 – 11.* Cambridge University Press.
Townsend, C. B. (1990). *World's toughest puzzles.* Sterling Publishing Company.
Townsend, C. B. (1994). *World's Most Amazing Puzzles.* Sterling Publishing Company.
Van Delft, P. & Botermans, J. (1978). *Creative puzzles of the world.* Cassell.
Vogt, D. (1994). *Streichholz Spiele. Eine kleine Geometrie des Vergnügens Gesucht und erfunden.* Insel Verlag.
Ward, A. (1986). *Help Your Child Creative Thinking.* Hodder and Stoughton Educational.
Wootton, S. & Horne, T. (2012). *Advanced brain training.* Hodder Education.

Chapter 8
An Unknown History of a Popular Matchstick Puzzle and its Changes Over Time

Abstract The aim of this chapter is to present an unknown history of the origin and later spread of a popular matchsticks puzzle: Make a polygon from 12 matchsticks with an area of 4 "matchstick squares". It was first published by Sam Loyd, Jr. in the Year 1925 in the newspaper "*The Atlanta Constitution*". In the Year 1934, the puzzle appeared in Soviet Union, published in Russian by Yakov Perelman. Textual details and the presentation of solution suggest that Perelman didn't invent the puzzle but "borrowed" it from Sam Loyd, Jr. In the Year1954, Boris Kordemsky largely reformulated Perelman's puzzle, eliminating introductory figure and reducing the sought area from 4 to 3 "matchstick squares". Puzzle's big popularity is mainly due to Martin Gardner who published the puzzle in the Year 1957 in the journal "*Scientific American*" and repeated it in his many books. There was a chance that Gardner could see the puzzle in Perelman's book "*Figures for fun*" that was published in English in January of the same year. Nobody of posterior authors gave credit to any one of previous authors, an unfortunate practice in the world of matchstick puzzles. A posterior reformulation of the puzzle, with the sought area reduced to two "matchstick squares", was suggested to make finding of its solution more difficult. The chapter is closed by describing students' creative solution of that highly challenging reformulation.

8.1 The First Publication of Puzzle in English

The puzzle was invented and published in the Year 1925 by Sam Loyd, Jr. (1873–1934). Sam was son of American "puzzle king" Sam Loyd (1871–1911). Following father's steps, he became an outstanding creator of puzzles. By his own estimation, Sam Loyd, Jr. created about 12,000 puzzles!

He was collaborator in many journals: conducted "Our puzzle school" in agriculture periodical "*Farm and fireside*", contributed by puzzles to "*Daily Eagle*" and "*Railroad man's Magazine*", a had his "Puzzle page" in "*Woman's Home Companion*".

J. Slisko, *Fostering Cognitive Mathematics Skills with Matchstick Puzzles*, Mathematics in Mind, https://doi.org/10.1007/978-3-032-14561-1_8

Two days after Sam Loyd Jr. died on February 23, "The New York Times" published a long obituary.

The puzzle of Sam Loy, Jr. appeared on Sunday Morning (May 17, 1925) in the Red Section of "*The Atlanta Constitution*". The Red Section was edited by "The world-famous Houdini" under the slogan "Education as well as Entertainment" (Loyd 1925a). The puzzle had the form presented in Fig. 8.1.

The answer to "the puzzling question" was published on the next Sunday Morning (May 24, 2025) in the same Red Section of "The Atlanta Constitution" (Loyd 1925b). The answer had the form given in Fig. 8.2.

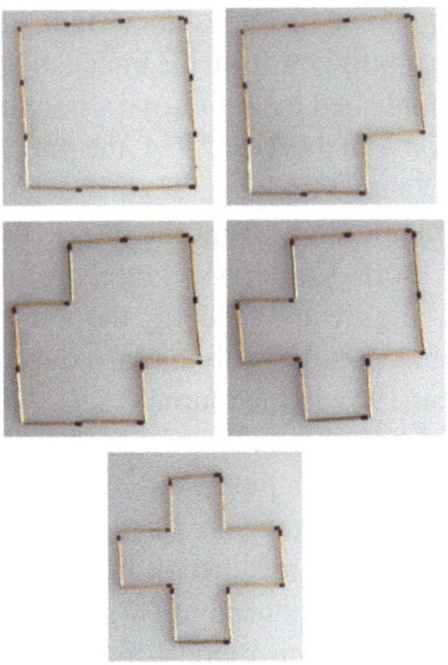

"Here is a problem in surveying that is well illustrated with match sticks.
Let us assume that each stick is an inch long. Then in the first figure – a square – we see how the twelve sticks may be arranged to enclose 9 square inches. In the second figure, the twelve sticks are arranged to enclose 8 square inches. In the third figure, they are made to enclose 7 square inches. In the fourth figure 6 square inches are enclosed.
Finally, in the fifth formation, only 5 square inches are enclosed.
Now comes the puzzling question: *How can the 12 match sticks be arranged in a continuous fence, 12 match sticks in length, so that the enclosed area shall be equal to 4 square inches, assuming that each stick is one inch in length.*
It is a perfectly fair problem, without trick of any sort, and *no measuring instruments are necessary* other than what measuring may be done with the sticks themselves."

Fig. 8.1 Text and figures of the puzzle as it was published by Sam Loyd, Jr. (*Source* Author).

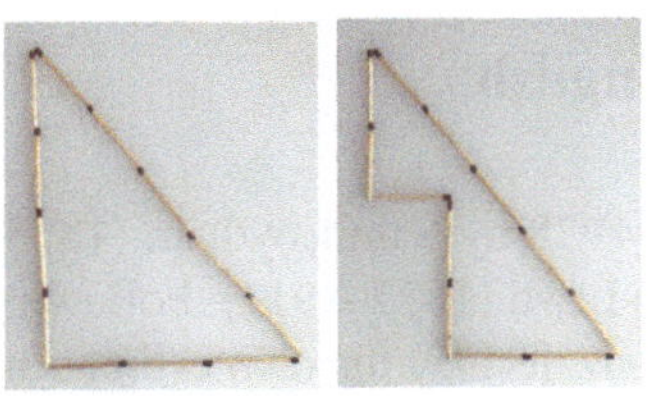

"Make the 12 matches into a triangle, as shown on the left, with 3 on one side, 4 on another, and 5 on the third. By the laws of geometry, this is a right triangle. Its area is one-half of 3 x 4, or six square inches. Now by moving only 3 of the matches, as show on the right, we subtract an area of exactly two square inches, leaving what we want."

Fig. 8.2 Text and figures of the answer to the puzzle as it was published by Sam Loyd, Jr. (*Source* Author).

8.2 The First Publication of Puzzle in Russian

The first publication of puzzle in Russian was done in the Year 1934 by famous Soviet science and mathematics popularizer Yakov Isidorovich Perelman (1882–1942). It was included in his book "*Alive mathematics*" (Perelman 1934). Before that, Perelman has already shown interest in matchstick puzzles, dedicating to them a whole book "*Fun with matchsticks*" (Perelman 1926). The puzzles included were mainly overtaken from Russian translation of Tromholt's book "*Matchstick games*". Perelman only invented a few puzzles in which the tasks were to construct different geometrical figures from a given number of matchstiks.

To the puzzle of Sam Loyd, Jr. Perelman gave the name "Out of 12 matchsticks". Its formulation (Perelman 1934, p. 150) is given in Fig. 8.3.

Text and supporting figures of Perelman's solution (Perelman 1934, pp. 157–158) are presented in Fig. 8.4.

Comparing the texts of both puzzle versions ("*no measuring instruments are necessary*" and "*use of measuring instruments is forbidden*", along with the figures and textual structure of both solutions, it would be hard to conclude that Perelman didn't know the puzzle published previously by Sam Loyd, Jr. As Perelman previously published in Russian many puzzles of Sam Loyd and Henry Ernest Dudeney, it seems that he had his own channels to get access to the puzzle literature in English language.

Fifteen years after Perelman's death, in January of 1957, the puzzle was included in the book "*Figures for fun. Stories and conundrums*". It was a collection of selected Perelman's recreational mathematical problems from various books, translated into English (Perelman 1957, p. 111, pp. 116–117).

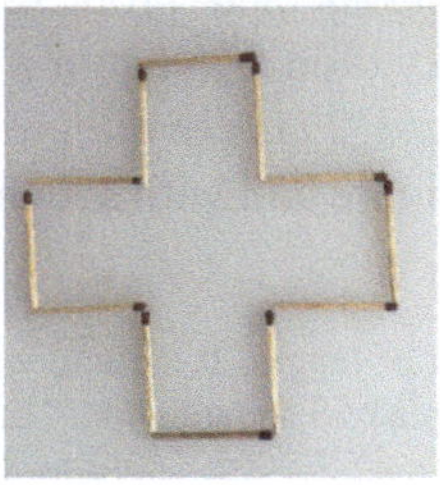

Out of 12 matchsticks you can build the figure of cross, with the area equal five "matchstick squares". Rearrange the configuration of matchsticks in such a way as that the figure's contour enclose an area equal to only four "matchstick squares". *The use of measuring instruments is forbidden.*

Fig. 8.3 Text and figure of Perelman's version of the puzzle previously published by Sam Loyd, Jr. (*Source* Author).

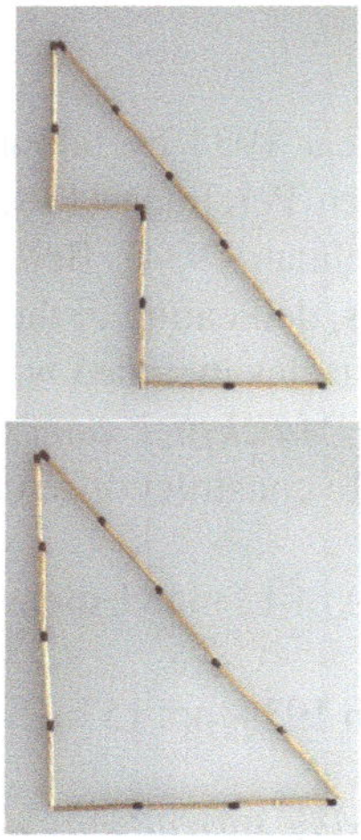

The matchstick should be laid out as shown in (the left) Figure. The area of that figure is equal to the quadrupled area of a "matchstick square". How is possible to be convinced of it? Let us mentally complete our figure to form a triangle. We get a right triangle whose base is equal to three and its altitude to four matchsticks. Its area is equal to one half of its base times its altitude: ½ x 3 x 4 = 6 squares whose sides are one "matchstick". But the area of our figure is obviously smaller than the area of the triangle by two "matchstick squares" and is, consequently, equal to four such squares.

Fig. 8.4 Text and figures of Perelman's solution to the puzzle previously published by Sam Loyd, Jr. (*Source* Author).

8.3 The Second Publication of Puzzle in Russian

The second publication of the puzzle "12 matchsticks make a polygon with the area of 4 square units", although with substantial changes, appeared in the Year 1954. Boris Anastasyevich Kordemsky (1907–1999) included it the first edition of his book "*Mathematical ingenuity*" (Kordemsky 1954).

Kordemsky was a recognized mathematics educator, well known for his many mathematical puzzles. Kordemsky knew Perelman's books and problems and mentioned him in the Foreword. Nevertheless, Kordemsky didn't give him a credit for the puzzle. It might be that he considered that his puzzle "*Find the outline of the figure*" is very different, without having an introductory figure and with sought area reduced to 3 square units (Kordemsky 1954, p. 99):

> There are 12 matchsticks. Let's accept every one of the as unit of length. It is required to form with 12 matchsticks such a figure which would close the area of 3 square units. This is a difficult task.

The text and figure of Kordemsky's solution (Kordemsky 1954, p. 465) are given in Fig. 8.5.

In the Foreword of the second edition of the book "*Mathematical ingenuity*" (Kordemsky 1955), Kordemsky said that some readers, who were independently solving the tasks, found, in some cases, additional or simpler solution and kindly sent to him their results. Kordemsky promised that the most interesting solutions will be mentioned in corresponding places in the book.

Keeping his word, for the puzzle "*Find the outline of the figure*", Kordemsky added the second "ingenious" solution, sent to him by engineer N. I. Arzhanov

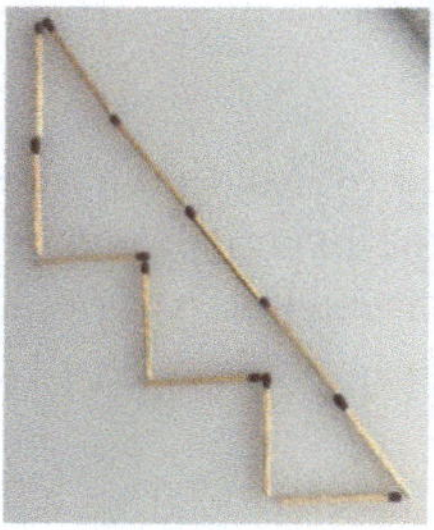

From 12 matchsticks one has initially to form a right-angle triangle with catheti of 3 and 4 matchsticks and hypothenuse of 5 matchsticks. The area of such triangle will be $\frac{1}{2} \cdot (3 \cdot 4) = 6$ square units.
If after that, the four matchsticks, making the right angle, are shifted in the form of ladder (as shown in Figure), area of triangle would decrease by 3 square units. Obtained figure… would be equal 6 – 3 = 3 square units.

Fig. 8.5 Text and figures of Kordemsky's solution to the puzzle "Find the outline of the figure" (*Source* Author).

from Leningrad. This solution with Kordemsky's description (Kordemsky 1955, pp. 440–441) is given in Fig. 8.6.

In the eighth edition (Kordemsky 1965, pp. 446–447), the puzzle had a new "ingenious" solution, sent to Kordemsky by engineer V. I. Lebedev from Moscow. The figure (with added matchsticks to make the construction clearer) and Kordemsky's description of this third solution are in Fig. 8.7.

Kordemsky's book was translated into English in Year 1972 with a changed title "*The Moscow puzzles. 359 mathematical recreations*". The puzzle in question had a slightly different name "Outline of a polygon" (Kordemsky 1972, p. 58) but presented the same three solutions (Kordemsky 1972, pp. 228–229).

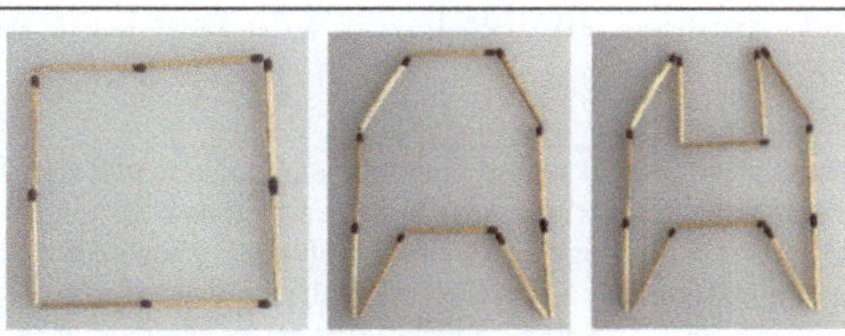

He formed a square with area of 4 square units, converted it in figure with equal area, after subtracted one square, obtained the figure, the area of which is 3 square units.

Fig. 8.6 Figures and Kordemsky's description of the second solution to the puzzle "Find the outline of the figure" (*Source* Author).

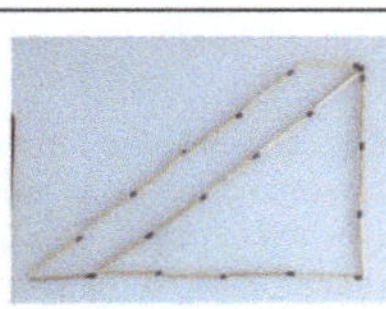

He formed a parallelogram with basis of one matchsticks and height of three matchsticks.

Fig. 8.7 Figure and Kordemsky's description of the third solution to the puzzle "Find the outline of the figure" (*Source* Author).

8.4 The Second Publication of Puzzle in English

The second publication of the puzzle in English was authored by Martin Gardner (1914–2010), the most famous mathematics popularizer in the twentieth century. He published the puzzle in November 1957 in his regular and very popular column "Mathematical games" in "*Scientific American*" (Gardner 1957a, 142, 144). Gardner's puzzle presentation is given in Fig. 8.8.

The solution (Gardner 1957b, p. 140) was published in December 1957 (Fig. 8.9).

As Perelman's book "*Figures for fun*" (Perelman 1957) was published in January of 1957, Gardner might have "borrowed" the puzzle from it, without giving him a credit as he usually did with other authors. If in that year Gardner didn't have Perelman's book, the other possibility is that Gardner saw the puzzle published by Sam Loyd, Jr. in the Year 1925.

In the Year 1961, Gardner definitively had Perelman's book and gave him a credit for the puzzle "*The repetitious number*" (Gardner 1961, pp. 54–55). In solution to this mathematical puzzle, Gardner wrote (Gardner 1961, p. 59): "This problem is given by Yakov Perelman in his book Figures for Fun, Moscow, 1957".

It is very strange that, after 1961, Gardner republished the puzzle in various of his books (Gardner 1994, p. 7, p. 45; Gardner 2006, p. 113, p. 128; Gardner 2008, p. 125, p. 132), without saying a word about the fact that Perelman had the same puzzle in the book "*Figures for fun*".

In the Year 1959, Gardner published the puzzle "*Twelve matches*" in the collection of his columns from "*Scientific American*" (Gardner 1959, p. 112, p. 113). After his solution (Gardner 1959, p. 120), Gardner announced that "there are hundreds of other solutions" because "each of the five tetrominoes (figures made with four squares)

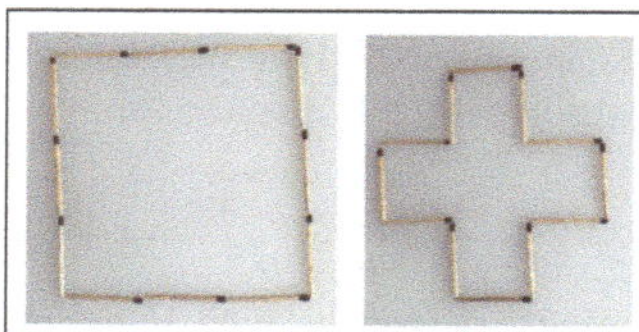

Assuming that a match is unit of length, it is possible to place 12 matches on a plane in various ways to form polygons with integral areas. On this page are two such polygons: a square with an area of nine square units, and a cross with an area of five. The problem is this: Use all 12 matches (the entire length of each match must be used) to in a similar fashion the perimeter of a polygon with an area of exactly four square units.

Fig. 8.8 Figures and text of Gardner's version of the puzzle (*Source* Author).

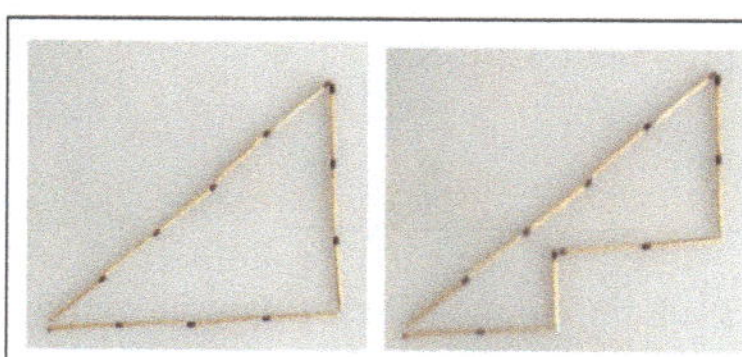

Twelve matches can be used to form a right triangle with sides of three, four and five units, as shown at left in the illustration at the top of page... This triangle will have an area of six square units. By altering position of three matches as shown at right in the illustration, we remove two square units, leaving a polygon with an area of four.

Fig. 8.9 Figures and text of Gardner's solution to the puzzle (*Source* Author).

can provide the base for a large number of solutions. We simply add and subtract the amount in triangular area to accommodate all twelve matches".

Gardner wrote that the correlation between tetrominoes and alternative solutions to the puzzle was suggested to him by "Elton M. Palmer, Oakmont, Pennsylvania". Gardner didn't reveal if Palmer was the author of 10 published representative solutions (two for every tetrominoe). Four solutions for the first two tetrominoes are in Fig. 8.10.

Following Kordemsky's practice to mention the profession of readers who sent original solutions, Gardner should have informed his readers that Elton M. Palmer was an engineer, employed at "Gulf Research & Development Company", Pittsburgh. In addition, Palmer was a highly successful inventor with various patents: "*Seismograph recording apparatus*" (Patented May. 4, 1948), "*Film transportation mechanism*" (Patented November 25, 1958) and "*Apparatus for profile presentation of seismic information*" (Patented August 4, 1959).

Using Palmer's approach, Bolt found fourth solution to Kordemsky's puzzle in which 12 matches should form polygon with the area of 3 "matchstick squares" (Bolt 1993, p. 94). Bolt's steps in constructing that solution are given in Fig. 8.11.

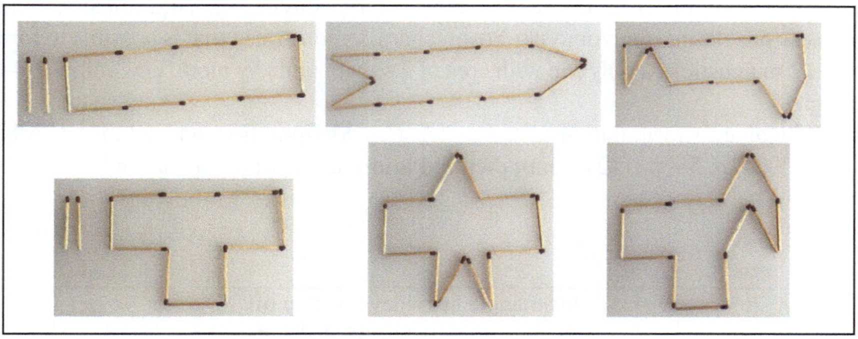

Fig. 8.10 Initial and final positions of 12 matchsticks in four alternative solutions based on the first two tetrominoes (*Source* Author).

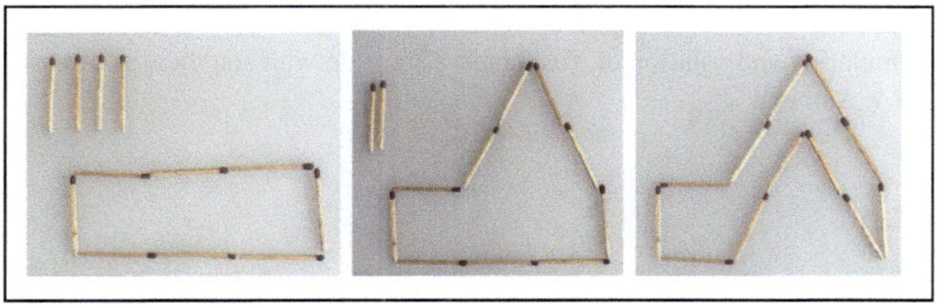

Fig. 8.11 Bolt's steps in constructing fourth solution to Kordemsky' puzzle (*Source* Author).

8.5 Puzzle with Reduced Sought Area to Two "Matchstick Squares"

As it was seen above, Kordemsky reformulated Perelman's puzzle reducing the sought area from four to three "matchstick squares". After Kordemsky's book appeared in English translation (Kordemsky 1972), it was likely an inspiration for Japanese puzzle-maker Fujimura to take the reformulations further.

Fujimura did it as follows. His puzzle "12/Reducing an area" (Fujimura 1978, p. 18) presented right-angle triangle made of 12 matchsticks and asked readers to find how to reduce its area from 6 squares units to 5 square units, to 4 square units and to 3 square units by moving 2 matchsticks, 2 more matchsticks and 2 more matchsticks.

His next puzzle "13/Two and one" reads (Fujimura 1978, p. 19):

> ... So far we have reduced the area of the figure to three square units, but we cannot reduce it further by the same procedure. Now try to obtain an area of two units by rearranging the matches to form a completely different figure. Finally, reduce the area to one unit.

Fujimura's solutions with parallelograms (Fujimura 1978, p. 118), having heights 1/2 and 1/4 of matchstick's length, aren´t very convincing. Additionally, other Japanese puzzle-master Nobuyuki Yoshigahara demonstrated that is possible to form from 12 matchsticks a polygon with area of 2 square units by modifying initial 3-4-5 triangle.

Yoshigahara's formulation, initial matchstick configuration, and solution (Yoshigahara 2004, p. 55, p. 112), contradicting Fujimura's affirmation, are in Fig. 8.12.

	This triangle has sides made up of three, four, and five matchsticks: its area is **(3x4)/2 = 6.** Move five matchsticks and create a shape whose area is two.	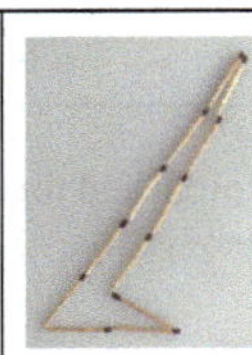

Fig. 8.12 Formulation and solution of Yoshigahara's puzzle with sought area of two "matchstick squares" (*Source* Author).

8.6 Students' Solutions to the Puzzles with Sought Area of Three and Two "Matchstick Squares"

With the aim to find out how self-regulated and collaborative learning foster students' creative thinking, Katz and Stupel (2015) designed and implemented in Israel an out-of-class, 2-hour activity with 24 six-grade students. The students were divided 12 pairs, each one having 12 matchsticks. The students' task was to explore various possibilities to create on their desks as many continuous polygons as they could. Each polygon had to have perimeter of 12 matchsticks in length and an area on integer number of "matchstick squares" (from nine to two). Students have shown an amazing creativity in finding asked polygons!

For Kordemsky's puzzle, students, using Palmer's approach, found Bolt's solution (Fig. 8.11) and three additional, fundamentally different solutions (Fig. 8.13)!

For Fujimura's puzzle "make from 12 matchsticks a figure with area of 2 square units" (Fujimura 1978, p. 19), students found, in two steps, a more elegant and easy-to-make solution! First, they formed from 6 matchsticks a parallelogram with an area of 2 "matchstick squares". After that, by moving 2 horizontal matchsticks and adding 6 matchsticks, they completed the sought figure with the same area (Fig. 8.14).

This exploration-oriented activity with matchsticks helped students to improve their creative abilities, self-efficacy, self-confidence, and communication skills. In addition, they changed their view on mathematics, recognizing its creative nature. Such important results suggest that matchstick puzzles should be used more

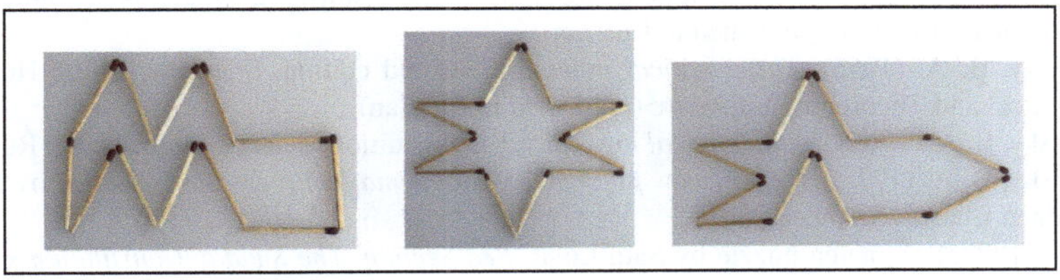

Fig. 8.13 Three never-published solutions to Kordemsky's puzzle found by students in Israel (*Source* Author).

Fig. 8.14 Students' solution to Fujimura's puzzle in two steps (*Source* Author)

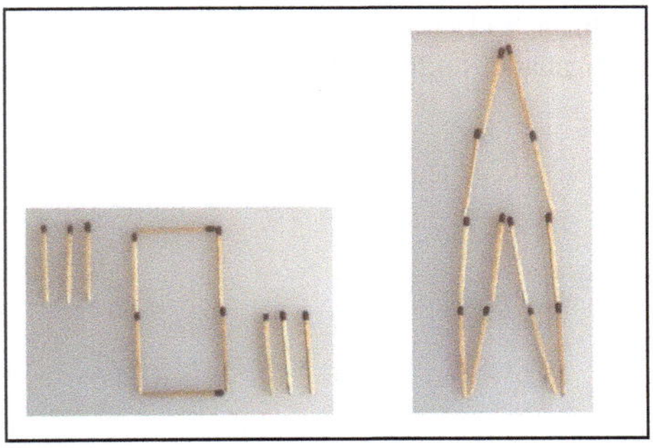

frequently in formal and informal education to foster cognitive mathematics skills of young and older puzzle solvers.

Surprising creativity of an anonymous reader of the fourth edition of Tromholt's book is presented in Chap. 9. More examples of students' impressive creativity in solving matchstick puzzles are given in Chap. 10.

References

Bolt, B. (1993). *A mathematical Pandora's box*. Cambridge University Press.

Fujimura, K. (1978). *The Tokyo puzzles*. Scribner.

Gardner, M. (1957a). Mathematical games. Nine titillating puzzles, the answer to which will be given next month. *Scientific American, 197*(5), 142, 144.

Gardner, M. (1957b). Mathematical games. *Scientific American, 197*(6), 140.

Gardner, M. (1959). *The Scientific American book of mathematical puzzles & diversions*. Simon and Schuster.

Gardner, M. (1961). *The 2nd Scientific American book of mathematical puzzles & diversions. A new selection*. Simon and Schuster.

Gardner, M. (1994). *My best mathematical and logic puzzles*. Dover Publications.

Gardner, M. (2006). *The colossal book of short puzzles and problems*. W.W. Norton and Company.

Gardner, M. (2008). *Hexaflexagons, probability paradoxes, and the Tower of Hanoi: Martin. Gardner's first book of mathematical puzzles and games*. Cambridge University Press.

Katz, S. & Stupel, M. (2015). Promoting Creativity and Self-efficacy of Elementary Students through a Collaborative Research Task in Mathematics: A Case Study. *Journal of Curriculum and Teaching*, 4(1), 68–82.

Kordemsky, B. A. (1954). *Mathematical ingenuity*. State Publishing House of Technical and Theoretical Literature (published in Russian).

Kordemsky, B. A. (1955). *Mathematical ingenuity*. Second edition. State Publishing House of Technical and Theoretical Literature (published in Russian).

Kordemsky, B. A. (1965). *Mathematical ingenuity*. Eighth edition. Science (published in Russian).

Kordemsky, B. A. (1972). *The Moscow puzzles. 359 mathematical recreations*. Edited by Martin Gardner. Charles Scribner's Sons.

Loyd, S. (1925a). A match puzzle by Sam Loyd. *Red Section. The Sunday Constitution's Young Folks Section. The Atlanta Constitution, 57*(338), 8.

Loyd, S. (1925b). Sam Loyd's Match Puzzle. *Red Section. The Sunday Constitution's Magazine.The Atlanta Constitution*, 57(345), 2.

Perelman, Y. I. (1926). *Fun with matchsticks*. Priboy (published in Russian).

Perelman, Y. I. (1934). *Alive mathematics*. ONTI (published in Russian).

Perelman, Y. (1957). *Figures for fun. Stories and conundrums*. Foreign Languages Publishing House.

Yoshigahara, N. (2004). *Puzzles 101. A puzzlemaster's challenge*. A. K. Peters.

Chapter 9
Amazing Creativity of an Anonymous Reader

Abstract This chapter presents the results of an uncommon case study of amazing creativity shown by an anonymous reader who has drawn by hand her or his solutions in the recent Anaconda's re-edition of the fourth edition of Tromholt's book "*Games with matchsticks*". After buying it as a second-hand book, I was in position to compare the reader's hand-drawn solutions with puzzle formulations and the solutions published by Tromholt and other book authors. For 13 puzzles, the reader was able to find Tromholt's solutions. More important is the fact that anonymous reader found different correct solutions to 12 puzzles. Three solutions were visually different and nine solutions were fundamentally different from the solutions published by Tromholt in the Anaconda's re-edition of the book.

9.1 Second-Hand Books as Possible Windows into Owners' Behavior

During many years I was a frequent buyer of second-hand books, starting long time ago with physics textbooks and recently looking for books about matchstick puzzles. Being so, I could observe different previous owners' behavior and treatment of the physics textbooks. Some kept the books as new, while other used different ways to mark the parts they considered important or the parts causing doubts. Many wrote short or long personal comments on those important parts.

In the case of books on matchstick puzzles I didn't observe such comments. The most of what some previous owners wrote were their personal names. From that, rather short and unimportant information, one can't conclude that they were or were not the readers of owned books. To my big surprise, the second-hand Tromholt's book "*Streichholzspiele*" (Fig. 9.1), published by Anaconda in 2007 (Tromholt 2007), was very different in two aspects, one negative and the other positive.

The negative aspect doesn't have anything to do with the previous owner. Namely, due for some strange reason, Kai Kilian decided to re-edit the fourth edition of Tromholt's book published in 1890 (Tromholt 1890), instead of the book's much improved fifth edition published in 1892 (Tromholt 1892).

J. Slisko, *Fostering Cognitive Mathematics Skills with Matchstick Puzzles*, Mathematics in Mind, https://doi.org/10.1007/978-3-032-14561-1_9

Fig. 9.1 Photo of the front page of the second-hand Tromholt's book (*Source* Author)

The positive aspect was: there were 28 pencil-drawn solutions previous anonymous owner left in my second-hand book. It showed that she or he was not only the owner but also a diligent reader and an amazingly creative matchstick puzzle solver.

The presence of hand-drawn solutions near puzzle formulations was an indication that the reader wanted to find her or his personal solutions while reading a particular puzzle. At the first sight, the presence of 28 solutions in the book with 285 puzzles might be considered as a very low score in puzzle solving. After a careful look at the solutions, I came to a different conclusion.

Those 28 hand-drawn solutions can be divided into three groups. In the first group are 3 wrong solutions. The second group included 13 solutions that are equal to the Tromolt's solutions given in the solution section. In the third and most important group, I found 12 solutions that are different from the solutions given by Tromholt. Three are *visually different* and nine are *fundamentally different* from Tromholt's solutions. In some puzzles, the reader's solutions were found neither by Tromholt in the fifth edition nor by later authors of books on matchstick puzzles. For this reason, I consider this anonymous reader had amazing creativity.

In this chapter I present and comment only on solutions belonging to the second and third group. Solutions by Tromholt will be named T-solutions and the hand-drawn solutions by the anonymous reader will be named A-solutions.

9.2 Puzzles with A-Solutions that Are Visually Different from T-Solutions

Below are two of three examples of such puzzles (Figs. 9.2 and 9.3).

1. Puzzle 3 (Tromholt 2007, p. 7, p. 109)

This solution is visually different from the solution given by Tromholt. Being a mirror image of Tromholt's solution, as previously discussed, it isn‘t considered fundamentally different. While some posterior book authors repeated Tromholt's solution (Lüders 1919, p. 40; Paraquin 1970, p. 26; Hansell 1981, p. 74; Anonymous 1996, p. 133; Picon 2002, p. 20, p. 115; Clarke 2017, p. 63), the solution of anonymous reader can be found published in the books written by other authors, both old

Two matchsticks are to be moved to create five squares of equal size.

Initial matchstick configuration | Matchstick configuration of T-solution

Fig. 9.2 Puzzle formulation in English, recreated initial, and solution configuration of Puzzle 3 (*Source* Author)

Hand-drawn A-solution | Recreated A-solution from matchsticks

Fig. 9.3 Hand-drawn and recreated solutions of Puzzle 3 by anonymous reader (*Source* Author)

(Czepa 1910, p. 61) and recent (Grund-Thorpe 2006, p. 73; Botermans 2006, p. 191; Anonymous 2014, p. 106; Anonymous 2017, p. 45, p. 112.

2. Puzzle 109 (Tromholt 2007, p. 35, p. 132)

Tromholt's solution and anonymous reader's solution are in Figs. 9.4 and 9.5.

This solution is considered visually different because it can be obtained from Tromholt's solution rotating it to the left by 90 degrees. A possible argument in favor of its originality is the following: While 40 of posterior book authors repeated Tromholt's solution or its mirror image, only 6 book authors published the solution equal to the solution found by anonymous reader.

Two matches are to be removed to create two squares.

Initial puzzle configuration | Matchstick configuration of T- solution

Fig. 9.4 Puzzle formulation in English, recreated initial, and solution configuration of Puzzle 109 (*Source* Author)

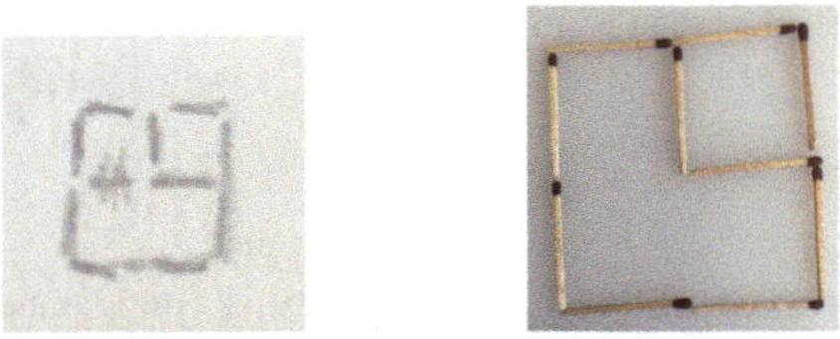

Hand-drawn A-solution Recreated A-solution from matchsticks

Fig. 9.5 Hand-drawn and recreated solutions of Puzzle 109 by anonymous reader (*Source* Author)

9.3 Puzzles with A-solutions that Are Fundamentally Different from T-Solutions

Below are 8 examples of such puzzles.

1. Puzzle 6a (Tromholt 2007, p. 9, p. 109)

Tromholt's solution and anonymous reader's solution are in Figs. 9.6 and 9.7.

It is easy to see that A-solution is fundamentally different from T-solution because it can't be obtained either as a rotation or as mirror version of T-solution. While some authors repeated Tromholt's solution (Botermans, 2006, p. 195), others published its rotation version (Paraquin 1970, p. 27; Anonymous 2014, p. 119; Clarke 2017, p. 52).

Two matchsticks are to be moved to create seven squares of equal size.

Initial puzzle configuration Matchstick configuration of T-solution

Fig. 9.6 Puzzle formulation in English, recreated initial and solution configuration of Puzzle 6 (*Source* Author)

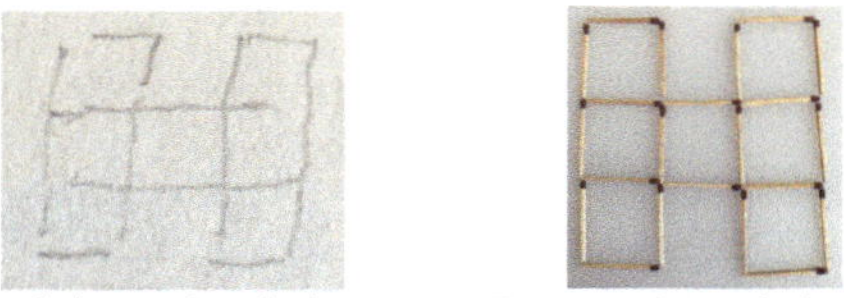

Hand-drawn A-solution Recreated A-solution from matchsticks

Fig. 9.7 Hand-drawn and recreated solutions of Puzzle 3 by anonymous reader (*Source* Author)

The hand-drawn solution of anonymous reader appeared in books on matchstick puzzles., too. Anonymous author of Tosa Verlag published an equal solution (Anonymous 1996, p. 155), while Paraquin published its rotation version (Paraquin 1970, p. 27).

2. Puzzle 19 (Tromholt 2007, p. 12, p. 113)

Tromholt's solution and anonymous reader's solution are in Figs. 9.8 and 9.9.

This A-solution is very important due to two reasons:

It is fundamentally different from T-solution.

It appeared in the fifth edition of Tromholt's book!

Namely, in that edition Tromholt reformulated the Puzzle 19 (Tromholt 1892, p. 14). In the first part of the puzzle, five squares were required to be equal and, in the second part, unequal. A-solution appeared as the solution for the first part and T-solution from the fourth edition was the solution for the second part (Tromholt 1892, p. 128).

Tromholt T-solution, with one 2 × 2 square and four 1 × 1 squares, appeared as the only solution for un-reformulated puzzle (Hansell 1981, p. 87).

3. Puzzle 22 (Tromholt 2007, p. 12, p. 114)

Tromholt's solution and anonymous reader's solution are in Figs. 9.10 and 9.11.

It is obvious that this A-solution is fundamentally different from T-solution and its repetitions in many puzzle books (Bakst 1954, p. 183; Paraquin 1970, p. 26; Brooke 1973; p. 22; Obermair 1977, p. 91; Holt 1978, p. 105; Brandreth 1982, p. 156; Dupré 1989, p. 179; Schuyt 1989, p. 143; Weaver 1992, p. 64; DiSpezio 1998, p. 90; Picon 2002, p. 134; Grund-Thorpe 2006, p. 80; Botermans 2006, p. 182).

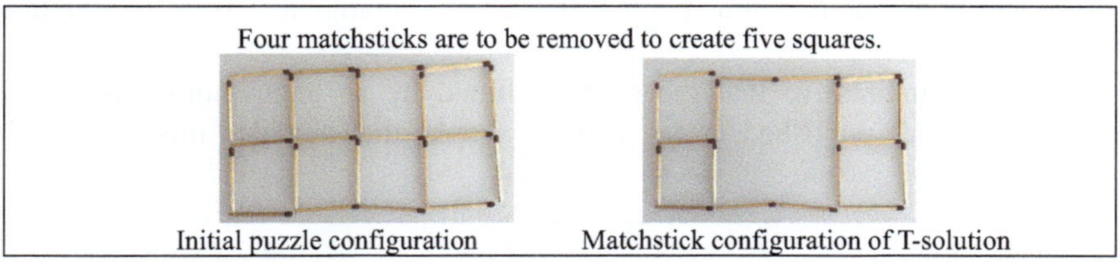

Fig. 9.8 Puzzle formulation in English, recreated initial and solution configuration of Puzzle 19 (*Source* Author)

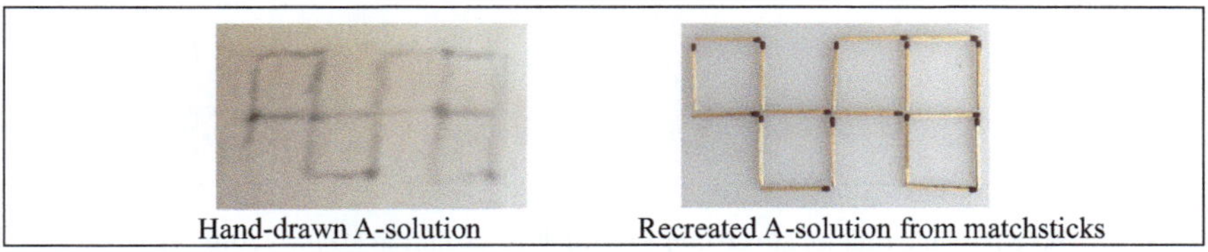

Fig. 9.9 Hand-drawn and recreated solutions of Puzzle 19 by anonymous reader (*Source* Author)

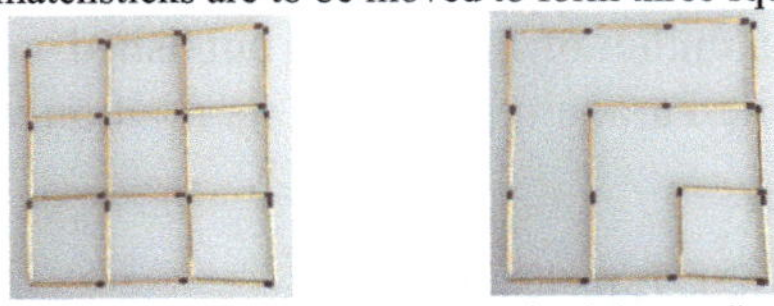

Fig. 9.10 Puzzle formulation in English, recreated initial and solution configuration of Puzzle 22 (*Source* Author)

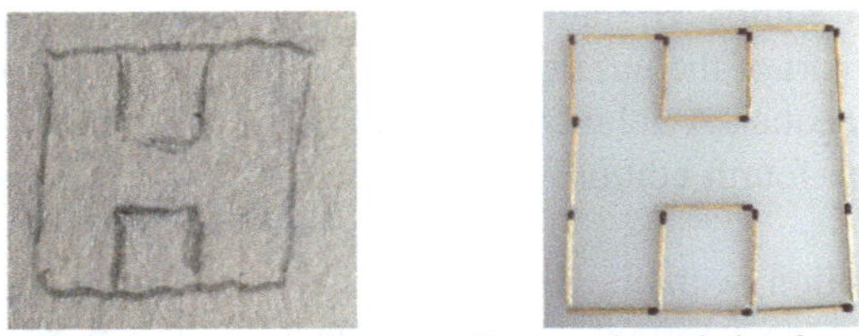

Fig. 9.11 Hand-drawn and recreated solutions of Puzzle 22 by anonymous reader (*Source* Author)

The A-solution of anonymous reader is also fundamentally different from three possible rotation versions of Tromholt's solution (Fig. 9.12).

The first rotation version was published by Uhl and Churchill (Uhl 1947, p. 39; Churchill 1985, p. 89).

The second rotation version was published by Eldin and two anonymous authors (Eldin 1982, p. 100; Anonymous 1996, p. 138; Anonymous 2017, p. 115).

Another anonymous author published the third rotational version (Anonymus 2014, p., 103).

It is quite impressive that the solution found by an anonymous reader (see Fig. 9.11) wasn't published in any of many books that included this Tromholt's

Fig. 9.12 Three possible rotational versions of Tromholt's solution (*Source* Author)

The 16 matchsticks must be rearranged so that they form 3 squares.

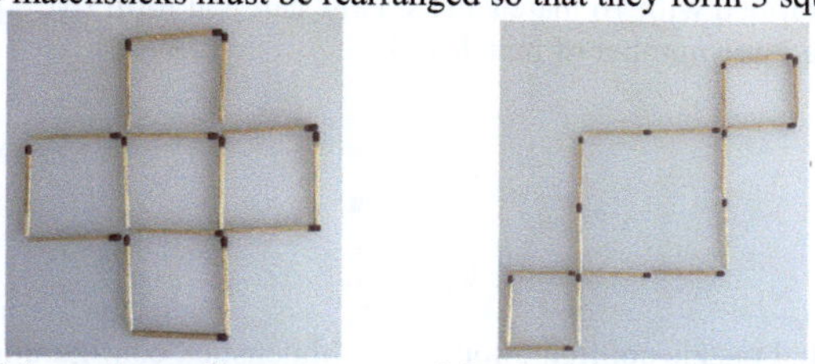

Initial puzzle configuration Matchstick configuration of T-solution

Fig. 9.13 Puzzle formulation in English, recreated initial, and solution configuration of Puzzle 93 (*Source* Author)

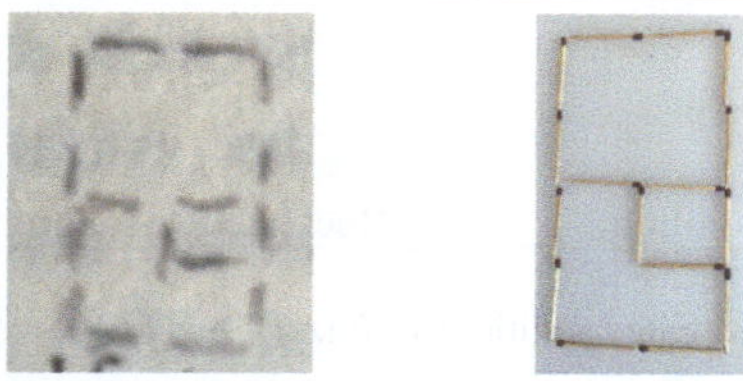

Hand-drawn A-solution Recreated A-solution from matchsticks

Fig. 9.14 Hand-drawn and recreated solutions of Puzzle 93 by anonymous reader (*Source* Author)

puzzle. Only one book published its rotation version (Andreescu, Cordero & Andreescu, 2020, p. 165)!

4. Puzzle 93 (Tromholt 2007, p. 31, p. 128)

Tromholt's solution and anonymous reader's solution are in Figs. 9.13 and 9.14.

This is another A-solution fundamentally different from T-solution. It differs also from the solution in the "second look" at this puzzle by Karl Heinz Paraquin, a very creative puzzle poser and solver. He added in his booklet (Paraquin 1970) quite a few additional solutions to Tromholt's puzzles.

The only change Paraquin did was to specify the number of matchstick moves one is allowed to carry out. While the number of moves was not specified in Tromholt's puzzle, in his puzzle 145, Paraquin asked that one must carry out just eight moves (Paraquin 1970, p. 23). Nevertheless, he only repeated Tromholt's solution (Paraquin 1970, 32).

5. Puzzle 99 (Tromholt 2007, p. 33, p. 130

Tromholt's solution and anonymous reader's solution are in Figs. 9.15 and 9.16.

A-solution is again fundamentally different from T-solution. This puzzle had also a "second look" by Paraquin, a better than one than in the previous puzzle. This time, in the Puzzle 27, Paraquin again specified the number of moves (Paraquing 1970, p. 4). If only 4 matchsticks should be moved, then Paraquin gets A-solution (Paraquin

Four squares are formed with 16 matchsticks. How can you form five squares of the same size with the same number of matchsticks.

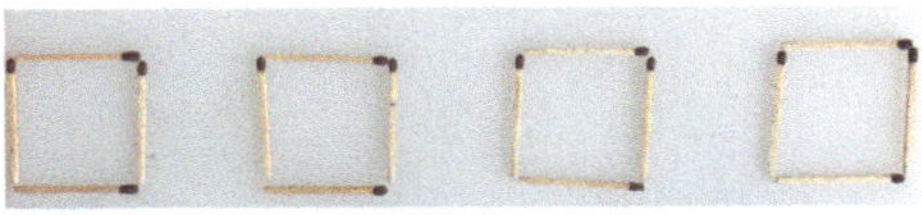

Initial puzzle configuration

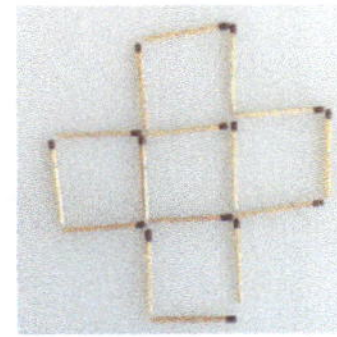

Matchstick configuration of T- solution

Fig. 9.15 Puzzle formulation in English, recreated initial, and solution configuration of Puzzle 99 (*Source* Author)

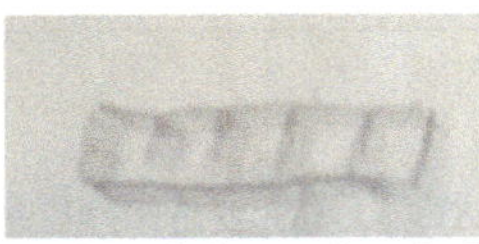

Hand-drawn A-solution

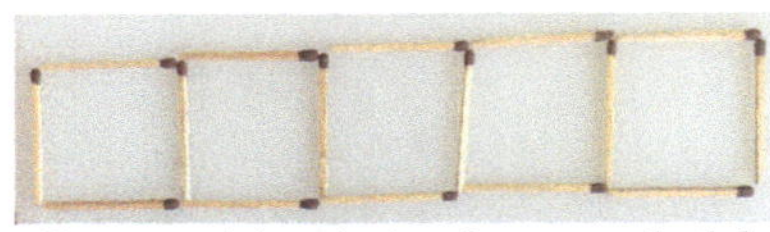

Recreated A-solution from matchsticks

Fig. 9.16 Hand-drawn and recreated solutions of Puzzle 99 by anonymous reader (*Source* Author)

1970, p 27). When 8 matchsticks should be moved, Paraquin got two solutions. The first one is fundamentally different from T-solution, while the second is Tromholt's solution (Paraquin 1970, p. 27).

6. Puzzle 130 (Tromholt 2007, p. 44, p. 135)

Tromholt's solution and anonymous reader's solution are in Figs. 9.17 and 9.18.

One time more, A-solution is fundamentally different from T-solution. It appeared two years later in the fifth edition of Tromholt's book (Tromholt 1892, p. 137)!

7. Puzzle 136a (Tromholt 2007, p. 45, p. 136)

Tromholt's solution and anonymous reader's solution are in Figs. 9.19 and 9.20.

Six matches are to be removed to create three squares.

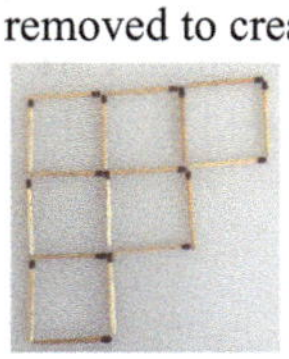

Initial puzzle configuration

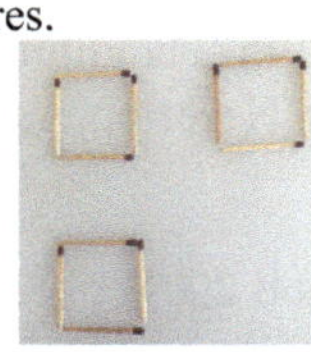

Matchstick configuration of T- solution

Fig. 9.17 Puzzle formulation in English, recreated initial, and solution configuration of Puzzle 130 (*Source* Author)

Hand-drawn A-solution Recreated A-solution from matchsticks

Fig. 9.18 Hand-drawn and recreated solutions of Puzzle 109 by anonymous reader (*Source* Author)

Three matches are to be removed to create four squares.

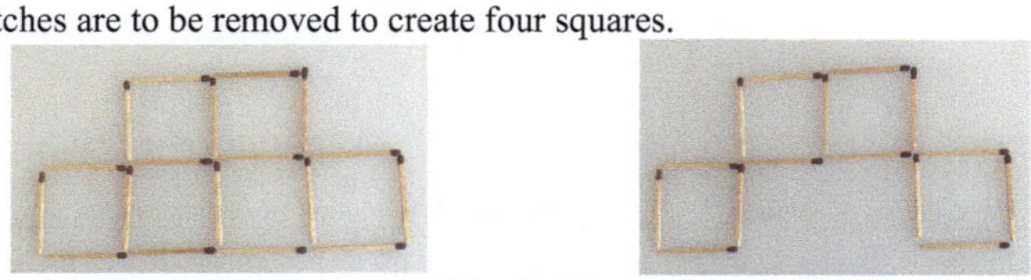

Initial puzzle configuration Matchstick configurations of two T- solutions

Fig. 9.19 Puzzle formulation in English, recreated initial, and T-solutions configuration of Puzzle 136a (*Source* Author)

Hand-drawn A-solution Recreated A-solution from matchsticks

Fig. 9.20 Hand-drawn and recreated solutions of Puzzle 109 by anonymous reader (*Source* Author)

This A-solution is also fundamentally different from T-solution and again appeared two years later in the fifth edition of Tromholt's book (Tromholt 1892, p. 138)!

8. Puzzle 136b (Tromholt 2007, p. 45, p. 136)

Tromholt's solution and anonymous reader's solution are in Figs. 9.21 and 9.22.

A-solution is again fundamentally different from T-solution and wasn't contained in the fifth edition of Tromholt's book. It was found much later by Paraquin (1970, p. 26)!

Six matches are to be removed to create three squares.

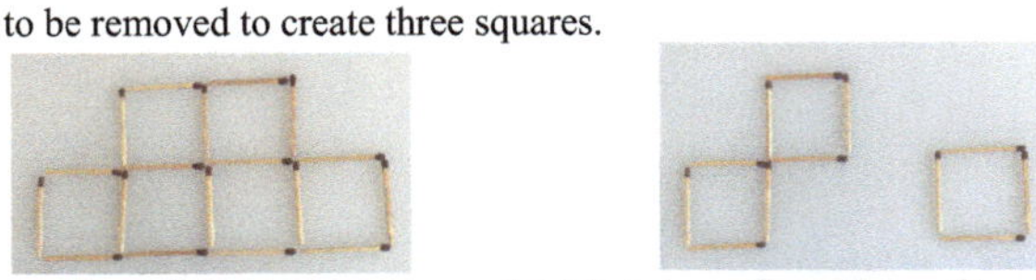

Initial puzzle configuration Matchstick configurations of two T- solutions

Fig. 9.21 Puzzle formulation in English, recreated initial, and T-solutions configuration of Puzzle 136b (*Source* Author)

Hand-drawn A-solution Recreated A-solution from matchsticks

Fig. 9.22 Hand-drawn and recreated solutions of Puzzle 136b by anonymous reader (*Source* Author)

9.4 A Final Question and a Possible Answer on A-Solutions

Amazing creativity shown by anonymous reader of the 2007 re-edition of Tromholt's book forced me to ask the following question: Why the reader didn't hand-draw more personal solutions?

A possible answer might be: "*Too many times my solutions differed from author's solutions. As author's solutions must be only correct ones, my solutions are likely wrong. It's better not to deal more with these puzzles!*".

This possible answer should be a warning call for book authors, asking them to foster their creativity to find all possible fundamentally different solutions for the puzzles they publish. Nobody has right to make a single person to feel less creative than she or he really is!

Considering the enormous popularity of Tromholt's book, in the late XIX century, it is possible that some of creative readers sent him the solutions he later included in the fifth edition, without giving them a well-deserved credit as Kordemsky and Gardner did.

References

Andreescu, T., Cordeiro, K. & Andreescu, A. (2020). *Awesome Math. Teaching mathematics with problem-based learning*. Jossey-Bass.

Anonymous (1996). *Streichholz-Spiele*. Tosa Verlag.

Anonymus (2014). *Das fabelhafte Büchlein der besten Streichholzspiele*. Regionalia Verlag.

Anonymous (2017). *Matchstick puzzles. Test your brain power with these challenging puzzles*. Arcturus.

Bakst, A. (1954). *Mathematical puzzles and pastimes*. Van Nostrand.
Botermans, J. (2006). *Matchstick puzzles. Over 200 of the world's most incredible puzzles*. Sterling Publishing Company.
Brandreth, G. D. (1982). *The complete puzzler*. St. Martin's Press.
Brooke, M. (1973). *Tricks, games and puzzles with matches*. Dover Publications.
Churchill, E. R. (1985). *Devilish bets to trick your friends*. Sterling Publishing Company.
Clarke, R. J. (2017). *Matchstick puzzles with answers*. CreateSpace Independent Publishing Platform.
Czepa, A. (1910). *Streichholz-Spielereien: Kunststücke und Scherze nebst Auflösungen*. Otto Maier Verlag.
DiSpezio, M. A. (1998). *Challenging critical thinking puzzles*. Sterling Publishing Company.
Dupré, J.P. (1989). *Junior fact finder: An illustrated encyclopedia for children*. Barron's.
Eldin, P. (1982). *Match Play. Safe puzzles, games and tricks with matchsticks*. Granada Publishing.
Grund-Thorpe, H. (2006). *Die schönsten Streichholzspiele.*Weltbild Verlag.
Hansell, S. (1981). *109 knifflige Streichholztricks*. Otto Maier Verlag.
Holt, M. (1978). *Math puzzles and games*. Volumes I & II. Walker and Company.
Lüders, C. E. (1919). *Der Streichholzkünstler in der Westentasche*. Verlag von G. Danner.
Obermair, G. (1977). *Matchstick puzzles, tricks and games*. Sterling Publication Company.
Paraquin, K. H. (1970). *Spiel und Spass mit Hölzchen und Köpfchen*. Otto Maier Verlag.
Picon, D. (2002). *Allumettes*. Mango.
Schuyt, M. (1989). *Phantastische Zündholzspiele*, DuMont Buchverlag.
Tromholt, S. (1889). *Streichholzspiele*. First edition. Otto Spamer Verlag.
Tromholt, S. (1890). *Streichholzspiele*. Fourth edition. Otto Spamer Verlag.
Tromholt, S. (1892). *Streichholzspiele*. Fifth edition. Otto Spamer Verlag.
Tromholt, S. (2007). *Streichholzspiele*. Anaconda.
Uhl, C. (Editor) (1947). *Der Kindergarten. Heft 8. Streichholzspiele*. Verlag Werden und Wirken.
Weaver, J. (1992). *Brain teasers*. Watermill Press.

Chapter 10
Stimulating Creativity by Information About the Number Possible Puzzle Solutions: A Few Small-Scale Pilot Studies

Abstract This chapter presents initial experimental evidence that the information about the number of possible solutions of a puzzle, given to puzzle solvers, stimulates their creativity. Related pilot studies were carried out with six geometric and two arithmetic matchstick puzzles for which many book authors commonly published only one solution, not escaping from the cognitive bias "one puzzle—one solution". Puzzle solvers involved were asked in different wordings to find multiple solutions, which are either visually or fundamentally different. They were informed about the number of sought solutions, supposing that this information will stimulate and increase their creative performances. In all studies, participants were more creative than book authors, finding more puzzle solutions, some of which were never published in the literature. These results strongly suggest that the authors of book with matchstick puzzles should include, after formulations of all multiple-solution puzzles, the number of possible different solutions and publish them in solution section.

10.1 The First Puzzle of Tromholt "Move 7 Matchsticks to Get 4 Squares"

I start presentations of the results of my small-scale studies with the first puzzle in Tromholt's book (Tromholt 1889, p. 1, p. 101; Tromholt 1892, p. 7, p. 127). Its formulation, (a) initial matchstick configuration, and (b) solution are given in Fig. 10.1.

This puzzle with only one symmetrical Tromholt's solution was repeated in many posterior books (Bakst 1954, p. 3, p. 183; Hansell 1981, p. 49, p. 76; Weaver 1992, p. 41, p. 63; Williams 2013, p. 11, p. 65; p. 463; Troshin 2022, p. 61, p. 323). Nevertheless, the first Tromholt's puzzle has many fundamentally different solutions.

This puzzle was given to a group of first-semester physics students (N = 33) in the course "*Development of complex thinking skills*" (August–December 2024), who didn't have a previous experience in solving matchstick puzzles. They were informed that the puzzle has *many different solutions*. Students' task was to find major number

J. Slisko, *Fostering Cognitive Mathematics Skills with Matchstick Puzzles*, Mathematics in Mind, https://doi.org/10.1007/978-3-032-14561-1_10

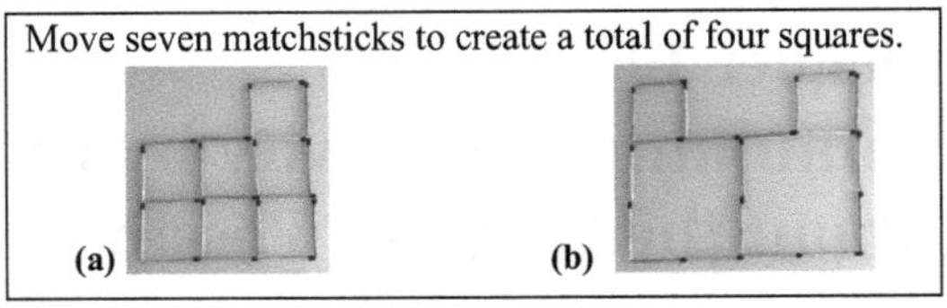

Fig. 10.1 Formulation, initial matchstick configuration, and solution to the first Tromholt's puzzle (*Source* Author)

of these solutions. An assistant in the course was taking photos of every found solution. The global results of this study will be presented in what follows.

Twenty students had poor performances. Eight of them didn't provide a solution. Twelve students only provided erroneous solutions. Seven of them one erroneous solution, three of them two erroneous solutions, and two of them three erroneous solutions.

Thirteen students were more successful. Three students provided only one correct solution, showing same, low level of puzzle solving creativity as Tromholt and posterior book authors. Puzzle solving creativity level of twelve students were higher: Five students provided *two correct solutions.* One student provided *three correct solutions.* One student provided *five correct solutions.* One student provided *nine correct answers.* One student provided *eighteen correct solutions* and one student provided *twenty correct solutions*!

Big majority of students' solutions belong to two different solution sequences (two 2 × 2 squares and two 1 × 1 squares) presented in Figs. 10.2 and 10.3.

To find all these solutions, one must have high skill of systematic visual thinking. Two students with one solution only found the first solution in the second sequence and one student with one solution only found the first solution in the first sequence. One student with three solutions only found eighth, nineth, and tenth solutions in the second solution sequence.

The highest skill level of systematic visual thinking was shown by students with the biggest number of solutions. The student with eighteen solutions found eight of twelve solutions in the first sequence and all ten solutions in the second sequence. The student with twenty solutions found eleven of twelve solutions in the first sequence

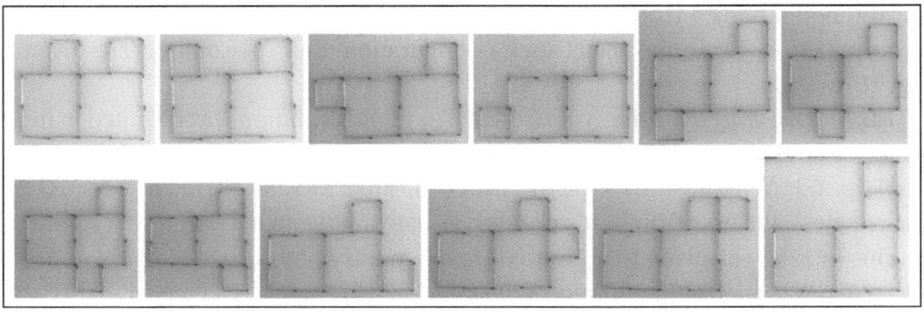

Fig. 10.2 The first solution sequence with two 2 × 2 squares and two 1 × 1 squares has twelve solutions. Tromholt's solution is the second one in this sequence (*Source* Author)

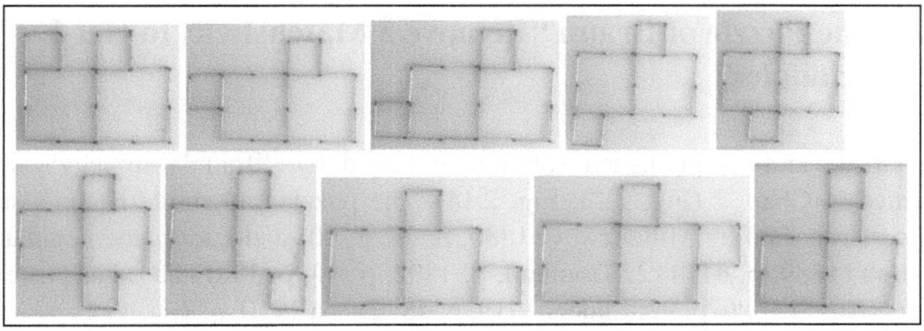

Fig. 10.3 The second solution sequence with two 2 × 2 squares and two 1 × 1 squares has 10 solutions (*Source* Author)

and nine of ten solutions in the second sequence. Only these two students found Tromholt's solution.

The biggest diversity in creativity was demonstrated by the student who found nine solutions in three different structural forms (Fig. 10.4).

In the first row there are all six solutions in the third solution sequence with two 2 × 2 and two 1 × 1 squares. In the second row there are two of six solutions in the fourth solution sequence with two 2 × 2 squares and two 1 × 1 squares. The nineth solution is the one of two solutions in the solution sequence with one 3 × 3 square and three 1 × 1 squares.

It is evident that twelve students in this study revealed more creativity while solving the puzzle in question than Tromholt and all posterior book authors. Knowledge that the puzzle had more than one solution helped them escape cognitive bias "one puzzle—one solution", stimulating their creative thinking and improving their solving performances (from two to twenty solutions).

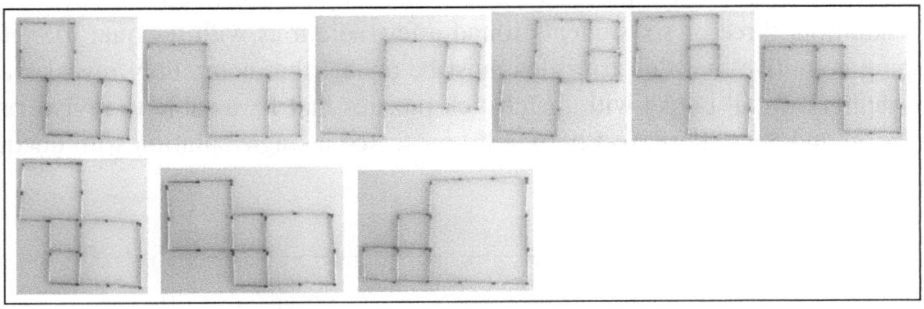

Fig. 10.4 Nine solutions of the students with biggest diversity in creativity (*Source* Author)

10.2 The Puzzle of Braun "Remove 5 Matchsticks to Get 5 Triangles"

This puzzle and its published solution, obtained by "Braun's strategy", were presented in Chap. 2 (Fig. 2.13; Fig. 2.14). The puzzle and single Braun's solution were repeated by Mittenzwey (1880, p. 38, p. 90) and later made popular by Tromholt (1889, p. 41, 112; Tromholt, p. 1892, p. 49, p. 138) and other posterior authors (Busch 1896, p. 92; Lüders 1919, p. 28; Czepa 1910, p. 2, p. 49).

Only three fundamentally different solutions to Braun's puzzle with five equal triangles were published in books with matchstick puzzles (Fig. 10.5).

The first two solutions appeared together in Russian translation of Tromholt's book (Tromholt 1912, p. 47, p. 134), showing a creative moment of an anonymous translator. These two solutions together (in reverse order) were published recently (Troshin 2022, p. 60, p. 324). The second solution as single one was published by various book authors (Obermair 1977; p. 34, p. 110; Picon 2002, p. 10, p. 104; Grund-Thorpe 2006, p. 31, p. 102). The third solution was published in only two books (Keiler and Michalski 1963, p. 138; Anonymous 2010, p. 45, p. 114).

A systematic analysis of possible solutions with five unequal or equal triangles, carried out by Heinrich Hemme, is given in the Appendix.

In a small-scale study, Braun's puzzle was given to six female students as a voluntary home activity via email. They were asked to find *at least three* fundamentally different solutions. The worksheet used in the study is presented in Fig. 10.6.

The results obtained in this study, with information about students' careers and short comments in figure descriptions, are in Figs. 10.7, 10.8, 10.9, 10.10, 10.11, and 10.12.

All students who participated in this small-scale study were creative puzzle solvers, being able to find for Braun's puzzle more different solutions than Tromholt and other posterior book authors who commonly only published one solution.

In addition, three of six students found a few solutions with unequal triangles, escaping cognitive bias "all triangles must be equal" that commonly reveal many book authors. In all books with matchstick puzzles that I was able to revise, only one (Scott and Scott 1978, p. 120, p. 186) presented a single solution with unequal triangles (Fig. 10.13).

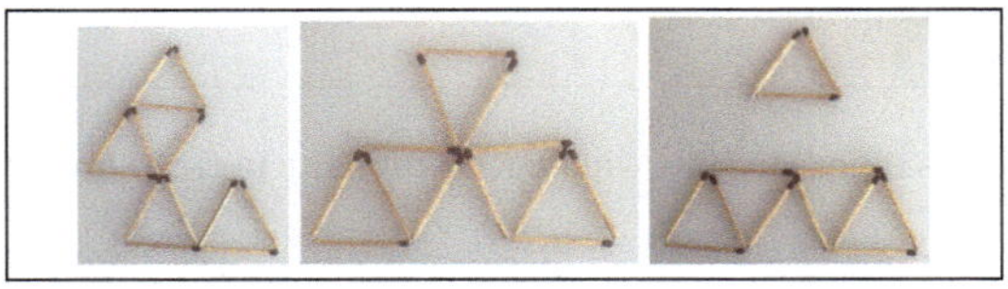

Fig. 10.5 Three fundamentally different solutions to Braun's puzzle with five equal triangles (*Source* Author)

The 1909 Nobel Prize winner in Physics, Ferdinand Braun, as a young physics and mathematics teacher, published in the Year 1876 one of the most popular matchstick puzzles. In the initial configuration (photo below), the matchsticks form nine equilateral triangles. The puzzle reads: Remove five matches to create five equilateral triangles.

Although this puzzle has many solutions, the vast majority of book and website authors publish only one solution. It's very rare for authors to publish two solutions.

Explore your mathematical creativity and visual intelligence and find at least three different solutions (between which there are no rotation or mirror relationships). In nine photos that follow, mark with an "X" the five matchsticks that must be removed to obtain a particular solution.

Fig. 10.6 Worksheet used in this study on students' creativity in solving Braun's puzzle with five triangles (*Source* Author)

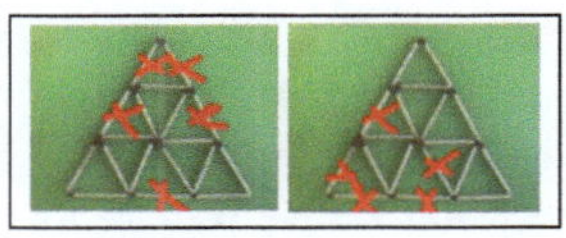

Fig. 10.7 Two solutions of María Fernanda (master student in mathematics education). She also gave two of their rotation variants (*Source* Author)

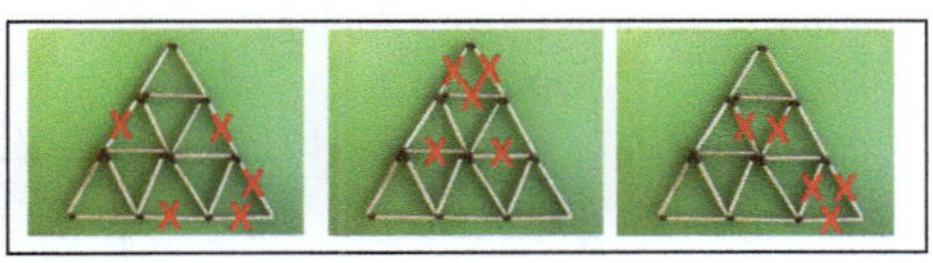

Fig. 10.8 Three solutions of Valeria (master student in mathematics education). *The second and the third solutions are with five unequal triangles* (*Source* Author)

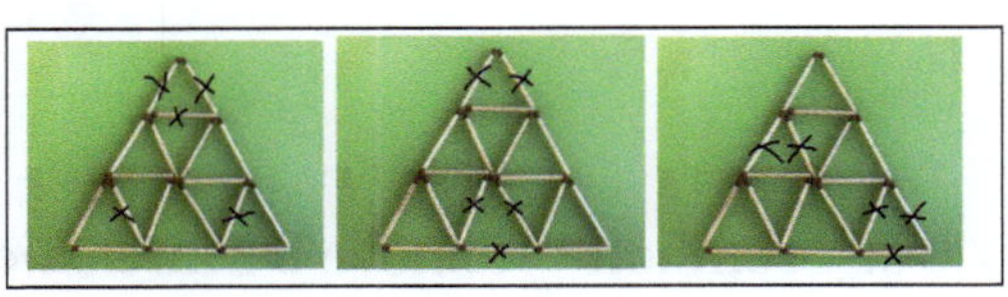

Fig. 10.9 Three solutions of Claudia (PhD student in mathematics education). She gave their six rotation variants, too. *The first solution is with five unequal triangles* (*Source* Author)

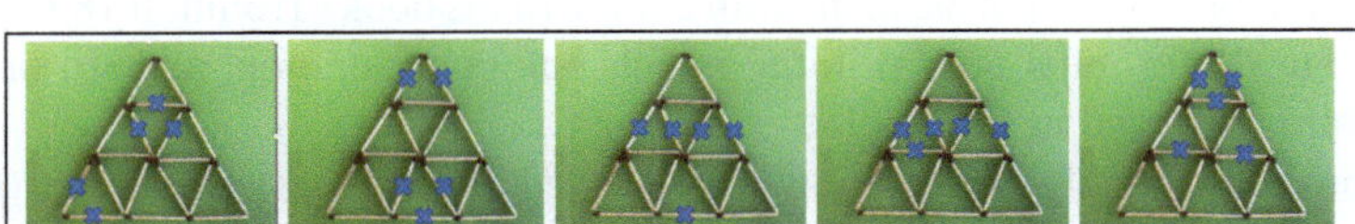

Fig. 10.10 Five solutions of Yuritzi (undergraduate mathematics student education). *The first and fifth solutions are with five unequal triangles* (*Source* Author)

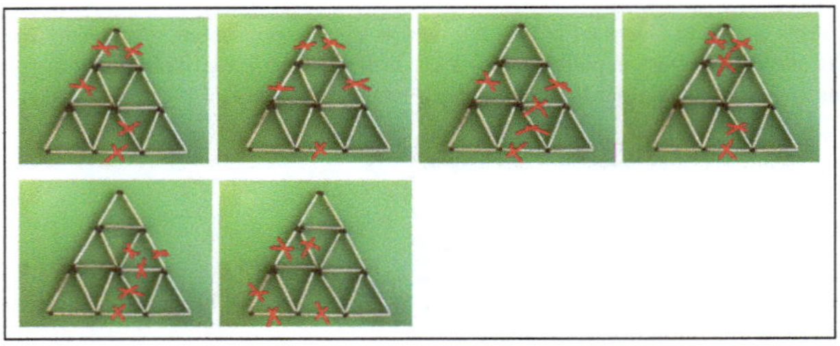

Fig. 10.11 Six solutions of Sofia (undergraduate physics student) (*Source* Author)

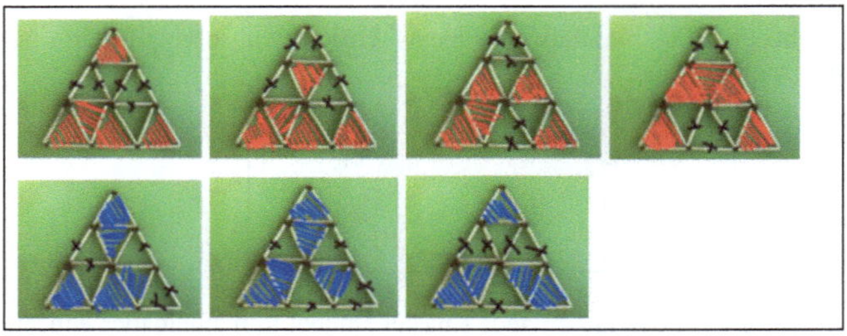

Fig. 10.12 Seven solutions of Alinne (master student in mathematics education). For four of her solutions she gave also their mirror versions (*Source* Author)

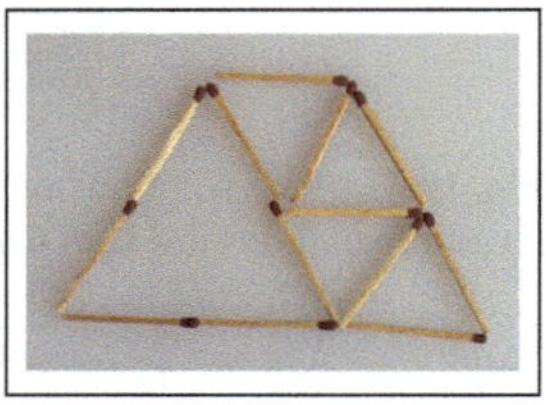

Fig. 10.13 Only one published solution to Braun's puzzle with five unequal triangles (*Source* Author)

10.3 The Puzzle of Tromholt "Remove 8 Matchsticks to Get 5 Equal Squares"

Tromholt published this puzzle in the fifth edition of his book (Tromholt 1892, p. 111, p. 147). Puzzle formulation, (a) initial matchstick configuration, and (b) only one solution are presented in Fig. 10.14.

The puzzle with only one solution was repeated in many posterior books (Lüders 1919, p. 45; Eicholz and O'Daffer 1969, p. 239; Brandreth 1981, p. 59, p. 231; p. 109; Stienecker 1997, p. 12, p. 28; Townsend 2002, p. 194, p. 627; Williams 2013, pp. 59–60, p. 137).

Only in one book (Parmeggiani and Santelia 1992, p. 243) appeared three fundamentally different solutions, whose existence was already mentioned after puzzle

"Fife equal squares should remain after removing… eight matchsticks."

(a)

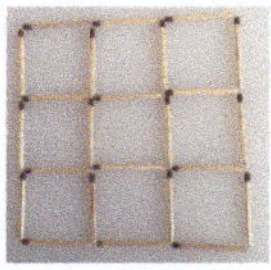

(b)

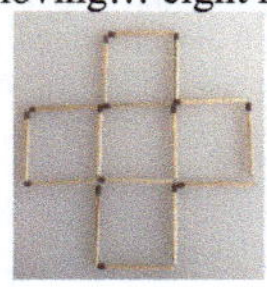

Fig. 10.14 Puzzle formulation, initial matchstick configuration, and solution of Tromholt's puzzle (*Source* Author)

formulation. The first was Tromholt's solution and two additional solutions were never published before. These solutions are presented in Fig. 10.15.

In December 2024 I found three additional, fundamentally different solutions to Tromholt's puzzle by considering possible variations of the second solution in Parmeggiani and Santelia's book (Fig. 10.16a). After sharing six solutions via email with Heinrich Hemme, he informed me that there was a seventh solution (Fig. 10.16b). His strategic approach to find seven fundamentally different solutions to Tromholt's puzzle is given in Appendix.

To explore age-related performances in solving this puzzle, a simultaneously two-part small-scale study was carried out. In the first part, the puzzle was given via email to a young, ten-year-old Norwegian boy Dennis for a mother-assisted home activity. Mother's assistance was necessary to explain to Dennis the concept of fundamentally different solutions. In the second part, the puzzle was given to five female students as a voluntary home activity via email. The worksheet, an editable Word document, used in the study is presented in Fig. 10.17.

Participants sent back their solutions in edited worksheet via email.

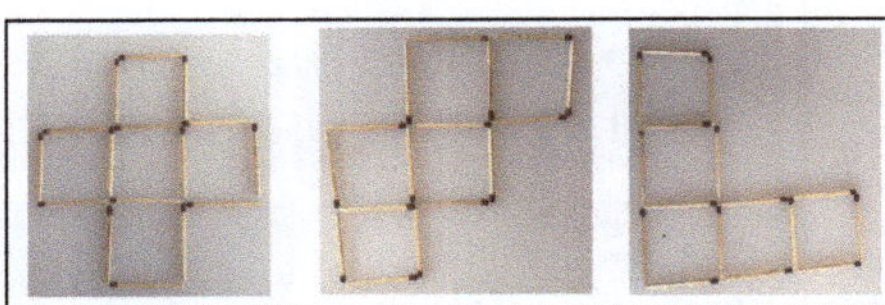

Fig. 10.15 Three fundamentally different solutions to Tromholt's puzzle published in Parmeggiani and Santelia 's book (*Source* Author)

(a)

(b)

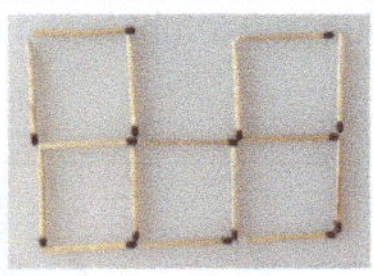

Fig. 10.16 Four additional, fundamentally different solutions to Tromholt's puzzle never published before (*Source* Author)

An interesting geometric puzzle with matchsticks

Using 24 matchsticks, you can form the following figure with squares:

The puzzle says: Remove 8 matchsticks to get 5 equal squares.

The interesting thing about this geometric matchstick puzzle is: The puzzle has 7 fundamentally different solutions. However, many book authors have only published one solution to this puzzle.

Activate your creativity and try to find all 7 fundamentally different solutions. For each new solution, check whether it can be obtained by rotation or as a mirror image of any of the previous solutions. If this is possible, then that new solution is not a fundamental solution and should not be included in the solutons.

Present your solutions in the seven photos below, indicating with the letter "X" in each one eight matchsticks that must be removed. All remaining matchsticks must be part of one of the 5 equal squares.

Fig. 10.17 Worksheet used in this study on creativity in solving Tromholt's puzzle with five equal squares (*Source* Author)

Dennis found in 40 min required seven solutions, not by thinking about them over worksheet photos but by manipulating matchsticks. As sent photos of Dennis' solutions had photographer's shadow, I asked Dennis' mother to send new photos. In the second solution making, Dennis found eighth fundamentally different solution! The photos of sitting-and-thinking Dennis and his eight solutions are given in Fig. 10.18.

Vannessa, undergraduate student of mathematics, like Dennis considered easier for her to manipulate matchsticks to find solutions and make own photos than to look for them by putting the letter "X" on offered seven photos in the worksheet. The photos of her five correct solutions are given in Fig. 10.19.

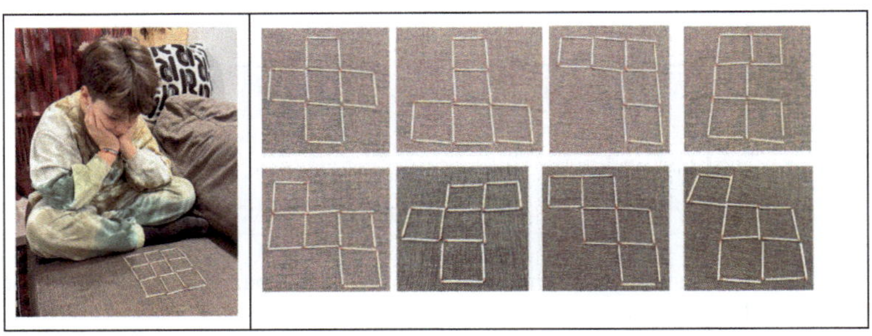

Fig. 10.18 The photos of sitting-and-thinking Dennis and his eight correct solutions. The first one is the solution of Tromholt and the last one is unknown-to-me solution (*Source* Author)

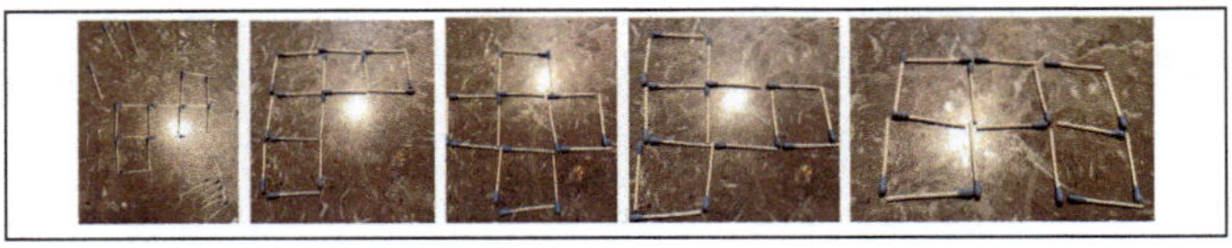

Fig. 10.19 The photos of Vannessa's five correct solutions. The third one is the solution of Tromholt's (*Source* Author)

As the eighth solution was unknown to me in the moment of designing the worksheet, asking participants to find only seven solution was quite misleading. This wrong information very likely limited their creative thinking. Namely, three participants with seven solutions found among them unknown-to-me solution and, consequently, stopped their solving processes. It can be seen in Figs. 10.20, 10.21, and 10.22.

Only Sofia, undergraduate student of physics, didn't follow seven-solution instruction and sent back eight solutions! These solutions are in Fig. 10.23.

This small-scale study brings two important results: **(1)** Tromholt's solution seems easy to find (it was the first solution for four of six participants). Its combination with cognitive bias "one puzzle—one solution" explains why so many book

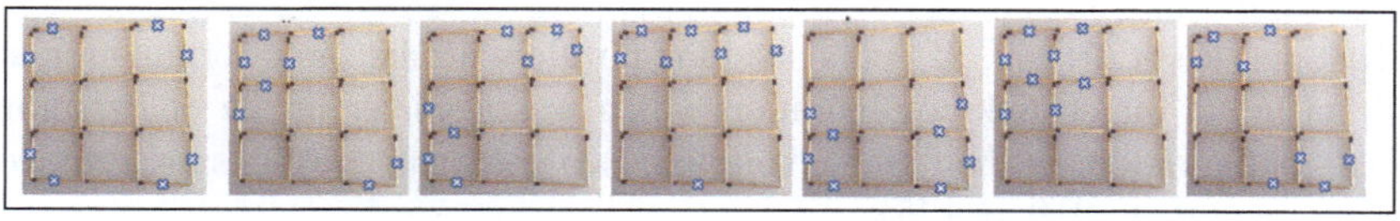

Fig. 10.20 Seven solutions of Yuritzi, undergraduate mathematics student. The first one is the solution of Tromholt and the third one is unknown-to-me solution (*Source* Author)

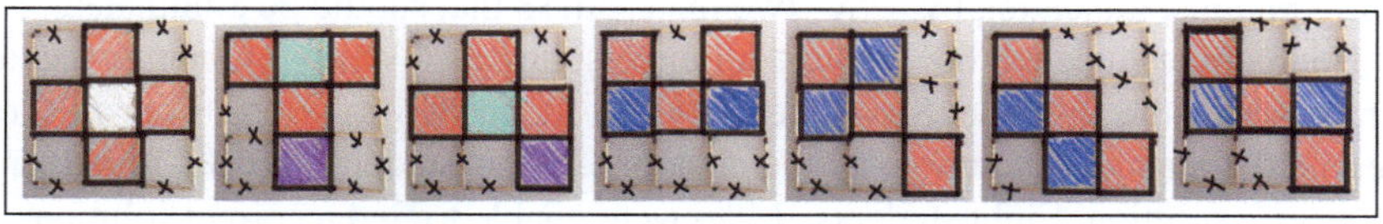

Fig. 10.21 Seven solutions of Alinne, master student in mathematics education. The first one is the solution of Tromholt and the fifth one is unknown-to-me solution (*Source* Author)

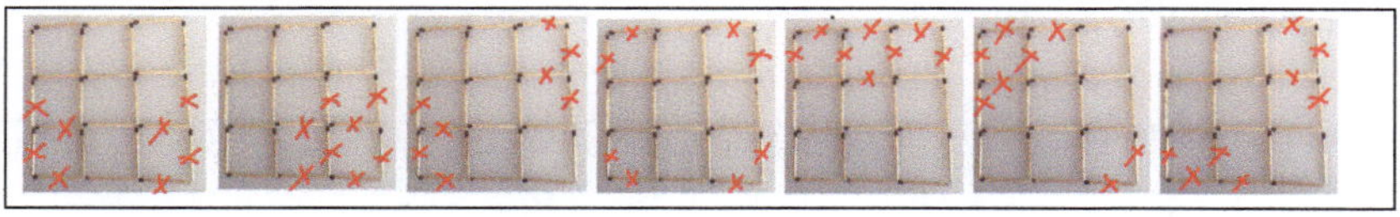

Fig. 10.22 Seven solutions of Claudia, PhD student in mathematics education. The fourth one is the solution of Tromholt and the seventh one is unknown-to-me solution (*Source* Author)

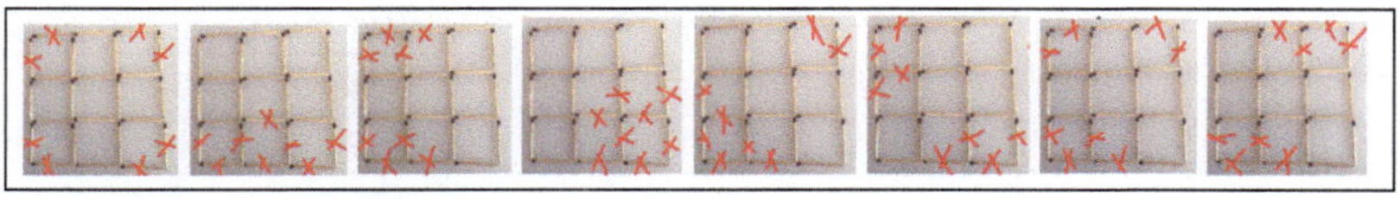

Fig. 10.23 Eight correct solutions of Sofia. The first one is the solution of Tromholt and the sixth one is unknown-to-me solution (*Source* Author)

authors repeated that solution. **(2)** Information about seven possible solutions stimulated participants' creativity and they performed much better than Tromholt and all posterior authors of puzzle books.

10.4 Paraquin's Puzzle "Remove Two Matchsticks to Convert 5 Squares into 4 Squares"

Formulation, (a) initial matchstick configuration and (b) solution of Paraquin's puzzle (Paraquin 1970, p. 2, p. 26) are given in Fig. 10.24.

Nevertheless, this puzzle has two additional, fundamentally different solutions, presented in Fig. 10.25.

Rist and Rist (2019, p. 35) published the puzzle with the same formulation and with an initial matchstick configuration which was a mirror image of Paraquin's initial configuration. In their single solution, two corner matchsticks are removed, too (Rist and Rist 2019, p. 125).

The puzzle was given to 56 Mexican middle-school students (ages between 13 and 15 years) as a paper-and-pencil test with information that the puzzle has three different solutions. Their task was to try to present them in three photos of initial matchstick configurations, indicating by a "X" two matchsticks that should be eliminated.

Thirty-three students (60%) found all three solutions. Twenty of them presented Paraquin's as the first one. Six students (11%) found two solutions. Four of them presented Paraquin's solution as the first one. Nine students (16%) found only Paraquin's solution. Eight students (14%) were unable to find a correct solution.

These experimental results show that information about the number of solutions made possible that many young students, without experience in solving matchstick puzzles, had better creative solving performance than book authors who included the that puzzle in their books.

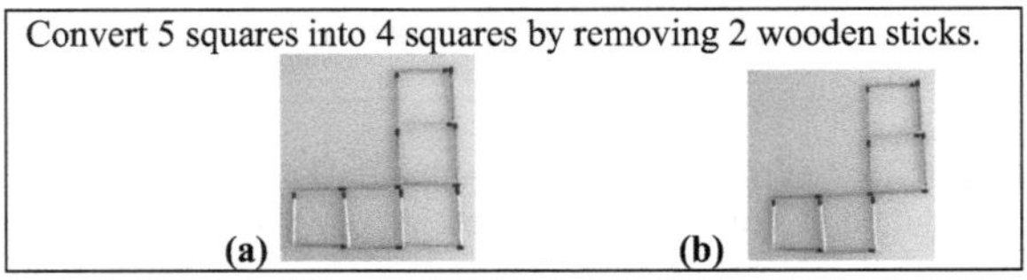

Fig. 10.24 Formulation, initial matchstick configuration, and solution to Paraquin's puzzle

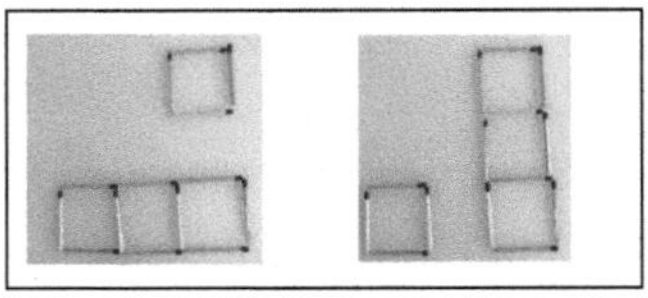

Fig. 10.25 Two additional, fundamentally different solutions to Paraquin's puzzle

10.5 Dudeney's Puzzle with Unrestricted Reformulation "Remove 4 Matchsticks to Get 4 Triangles"

Dudeney's puzzle "Remove 4 matchsticks to get 4 equal triangles" (Dudeney 1967, p. 206, p. 406) in its unrestricted reformulations "Remove 4 matchsticks to get 4 triangles" has a big number of possible solutions that includes various combinations of equal or unequal triangles. In a single-person study, the puzzle was given to Hrvoje, a 16-year-old high-school student in Croatia, with the aim to explore his creative thinking. The worksheet used for puzzle solving is Fig. 10.26.

Twelve solutions found by Hrvoje are given in Fig. 10.27.

The solutions 6 and 7 are with equal triangles and ten resting solutions contain unequal triangles. Hrvoje's solving creativity is amazing because, as it was seen in Chap. 7, for this puzzle book authors commonly published single solution with equal triangles. Additionally, only in one book there is a single solution with unequal triangles!

In the picture below, 16 matchsticks form 2 large and 8 small equilateral triangles.

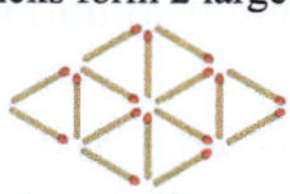

Remove 4 matchsticks so that only 4 equilateral triangles remain. There are many solutions!
Try to find at least 12 solutions. In the 12 pictures below, mark with an "X" the matchsticks that need to be removed.

Fig. 10.26 Worksheet used in this single-person study on creativity in solving Dudeney's puzzle in unrestricted reformulation (*Source* Author)

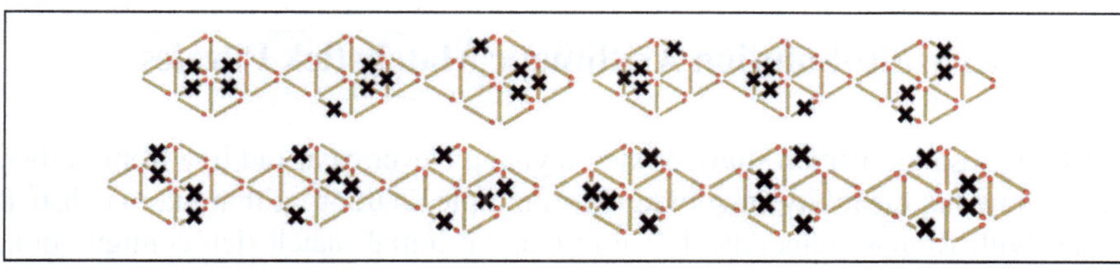

Fig. 10.27 Twelve solutions found by Hrvoje (*Source* Author)

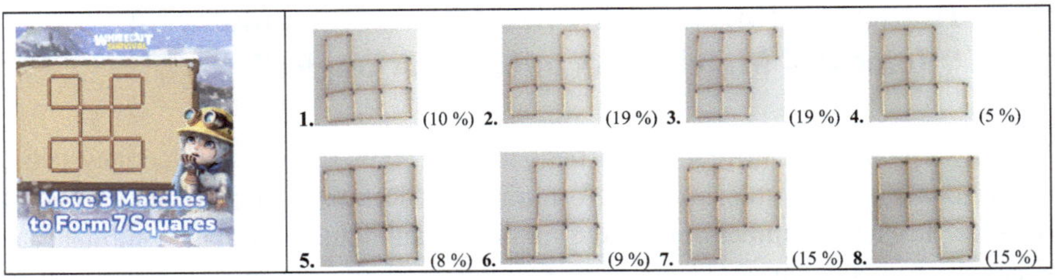

Fig. 10.28 The contest question and percentages of eight visually different solutions published by participants (*Source* Author)

10.6 Tromholt's Puzzle "Move 3 Matchsticks to Get 7 Squares": 8 Solutions Published in a Facebook's Contest

As it was mentioned in Chap. 6, this puzzle was a contest question on the wall of a popular Facebook site https://www.facebook.com/WhiteoutSurvival. The image of contest question and percentages of eight solutions published by 146 participants are given in Fig. 10.28.

The administrator of the Facebook site was accepting as correct only the solution 7 (Tromholt's solution). The reason might be that the administrator only knew the puzzle book in which that solution was published (Williams 2013, p. 111). Three solutions with the lowest percentages were solution 4, solution 5, and solution 6. As it was found in Chap. 6, these are those solutions of Tromholt's puzzle that didn't appear in any of published books. It seems that puzzle book authors don't find unlikely solutions!

10.7 Two Multi-Solution Arithmetic Matchstick Puzzles

In this micro single-person study, Viktor, a young eleven-year-old boy from Serbia, had to show his creativity in solving two multiple-solution arithmetic matchstick puzzles with Roman numerals. The formulation, initial matchstick configuration, and the solution of the first puzzle (Brooke 1973, p. 5, p. 40; Eldin 1982, p. 23, p. 109) are in Fig. 10.29.

Move one matchstick and make this equation valid.

Fig. 10.29 The formulation, initial matchstick configuration, and single solution of the first multi-solution puzzle (*Source* Author)

This puzzle was given to Viktor with the task to find *three different solutions*. His solutions and corresponding solving times are presented in Fig. 10.30.

The formulation, initial matchstick configuration, and the solution of the second puzzle (Rist and Rist 2019, p. 19, p. 111) are in Fig. 10.31.

This puzzle was given to Viktor with the task to find *four different solutions*. His solutions made with wooden sticks and corresponding solving times are presented in Fig. 10.32.

Viktor performed much better in solving these two multi-solution matchstick puzzles than book authors who published these puzzles with single solutions. Namely, he found all seven possible solutions to the puzzles while the authors of books were able to find only two!

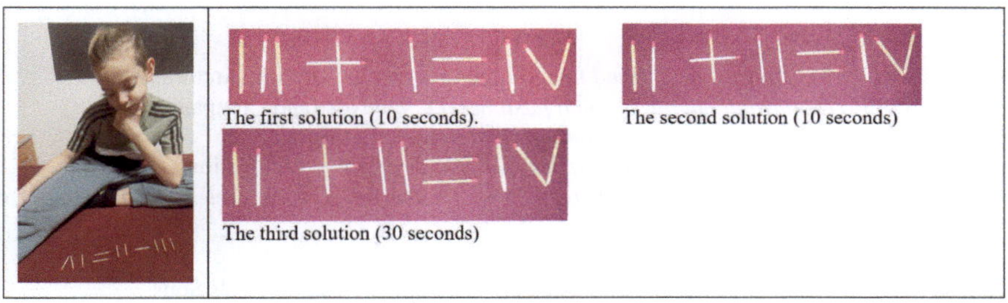

Fig. 10.30 Sitting-and-thinking Viktor and his three solutions to the first arithmetic matchstick puzzle found in 50 s (*Source* Author)

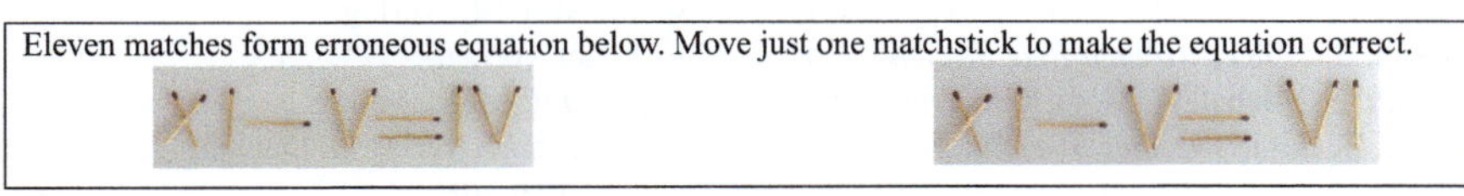

Fig. 10.31 The formulation, initial matchstick configuration, and solution of the second puzzle (*Source* Author)

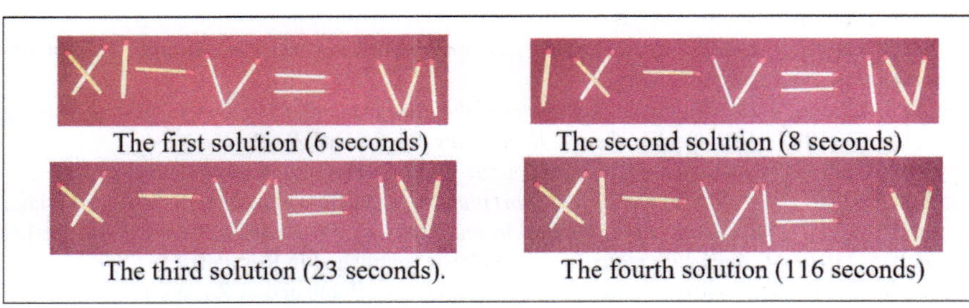

Fig. 10.32 Viktor's four solutions to the second arithmetic matchstick puzzle found in 153 s (*Source* Author)

10.8 Short Conclusion

The results of six small-scale studies, presented in this Chapter, are convincing initial experimental evidence that information about the number of possible puzzles' solutions stimulated solving creativity of the participants, majority of them without previous experiences with matchstick puzzles. Stimulated creativity made them able to find more solutions than all book authors. Additionally, these results show the falsehood of two beliefs: "all puzzles have only one solution" and "only some exceptional persons are creative"!

References

Anonymous (2010). *Brain teasers. Mind boggling pocket puzzles for kids.* Arcturus.
Bakst, A. (1954). *Mathematical puzzles and pastimes.* D. Van Nostrand Company.
Brandreth, G. D. (1981). *The complete home entertainer.* Robert Hale.
Brooke, M. (1973). *Tricks, games, and puzzles with matches.* Dover Publications.
Busch, A. (1896). *Die schönsten Gesellschaftspiele im Zimmer und im Freien.* August Schultze Verlag.
Czepa, A. (1910). *Streichholz-Spielereien.* Otto Maier Verlag.
Dudeney, H. E. (1967). 536 *puzzles and curious problems.* Charles Scribner's Sons.
Eicholz, R. E. & O'Daffer, P. G. (1969). *Elementary school mathematics.* Book 5. Teacher's edition. Second edition. Addison-Wesley.
Eldin, P. (1982). *Match Play. Safe Puzzles, Games and Tricks with Matchsticks.* Granada Publishing Limited.
Grund-Thorpe, H. (2006). *Die schönsten Streichholzspiele.* Weltbild Verlag.
Hansell, S. (1981). *109 knifflige Streichholztricks.* Otto Maier Verlag.
Keiler, I. & Michalski, M. (1963). *Zauberbuch für Kinder.* Otto Maier Verlag.
Lüders, C. E. (1919). *Der Streichholzkünstler in der Westentasche.* Verlag G. Danner,
Mittenzwey, L. (1880). *Mathematische Kurzeweil.* Julius Klinkhardt.
Obermair, G. (1977). *Matchstick puzzles, tricks and games.* Sterling Publishing Company.
Paraquin, K. H. (1970). *Spiel und Spass mit Hölzchen und Köpfchen.* Otto Maier Verlag.
Parmeggiani, T. & Santelia, C. E. (1992). *Il grande libro degli enigmi. Antologia di problemi insoliti, trappole logiche, e rompicapo de ogni tempo e latitudine*. Rizzoli Libri.
Picon, D. (2002). *Allumettes.* Mango.
Rist, M. & Rist, L. A. (2019). *Matchstick puzzles. Entertain your family and friends. 100 fun and challenging puzzles.* Penhagenco.
Scott, H. J. & Scott, L. (1978). *Close encounters of the third kind mind-benders.* Grosset & Dunlap.
Stienecker, D. (1997). *Discovering shapes. Rectangles.* Benchmark Books.
Townsend, C. B. (2002). *The world's biggest puzzle book.* Sterling Publishing Company.
Tromholt, s. (1889). *Sterichholzspiele. Denksport und Kurzweil.* First edition. Otto Spamer Verlag.3
Tromholt, s. (1892). *Sterichholzspiele. Denksport und Kurzweil.* Fifth edition. Otto Spamer Verlag.
Tromholt, S. (1912). *Games with matchsticks.* Mathesis (published in Russian).
Troshin, V. (2022). *Don't mess around with matches.* Litres. (Published in Russian).
Weaver, J. (1992). *Brain teasers.* Watermill Press.
Williams, B. (2013). *Gymnastics for the brain. Matchstick puzzles.* Salim Bouzekouk.

Chapter 11
Visual Intelligence in Finding Additional Solutions of Matchstick Puzzles

Abstract This chapter claims that analysis of a found puzzle solution, assisted by visual intelligence, can be used to find rather easily more additional visually or fundamentally different puzzle solutions. It is a complement to Braun's and Katona's strategies in which general geometrical features of solutions are found without specification about the positions of its parts. The chapter provides four examples which show how visually intelligence can help to find additional solutions from a published single one. Obviously, the authors, who published only one solution to many multiple-solution matchstick puzzles, failed to activate their visual intelligence.

11.1 Visual Intelligence: What Is and How to Foster it?

In 1983, to challenge a limited notion of intelligence, Howard Gardner introduced seven different types of intelligences: Linguistic, Logical/Mathematical, Spatial, Bodily/Kinesthetic, Musical, Interpersonal and Intrapersonal. Spatial intelligence was defined in the following way (Gardner 1983, p. 173):

> Central to spatial intelligence are the capacities to perceive the *visual world* accurately, to *perform transformations and modifications upon one's initial perceptions*, and to be able to re-create aspects of one's visual experience, even in the absence of relevant physical stimuli.

The followers of Gardner preferred to call this intelligence "spatial/visual intelligence" or simply "visual intelligence" For example, Weaver, in his considerations of "visual literacy", defines spatial/visual intelligence as an "ability to visualize objects, *move them around in space*" (Weaver 1999, p. 191). For Seward Barry, visual intelligence is "a quality of creative problem-solving that originates in perceptual process" (Seward Barry 1997, p. 8).

To foster visual intelligence, many authors generally recommend activities that include solving "visual puzzles" (Tamilselvi and Geetha 2015), "jigsaw puzzles" (Stine 2002, p. 14; Johnson 2010, p. 40) or "mazes" (Cathcart 1998). Surprisingly, nobody mentioned "matchstick puzzles" which are ideal tasks to practice and improve productive use of visual intelligence!

J. Slisko, *Fostering Cognitive Mathematics Skills with Matchstick Puzzles*, Mathematics in Mind, https://doi.org/10.1007/978-3-032-14561-1_11

In what follows, four examples of multiple-solutions matchstick puzzles are given for which authors only published one solution. For those puzzles, it is show how is possible, using visual intelligence, to get additional solutions, which are either visually or fundamentally different. It can be done by mentally moving some matchsticks to new positions while keeping required general structure of solution.

11.2 Remove 14 Matchsticks to get 6 Equal-Sized Squares

The first example is one of Sophus Tromholt's puzzles (Tromholt 1889, p. 6, p. 103; Tromholt 1890, p. 6, p. 106). Its formulation, initial matchstick configuration, and solution are given in Fig. 11.1.

In fifth edition of his book "Matchstick games", Tromholt claimed that this puzzle has two solutions (Tromholt 1892, p. 13) but only published one (Tromholt 1892, p. 128)!

Using visual intelligence, one gets easily 6 different solutions noticing that is possible to move four matchsticks of every square in the first row to two possible positions (Fig. 11.2).

Alois Czepa retook Tromholt's puzzle in slightly different formulation (Czepa 1910, p. 8): "Fourteen pieces of wood must be removed so that only six squares

14 matchsticks are to be removed so that 6 squares of equal size remain.

Fig. 11.1 Formulation and solution to Tromholt's puzzle (*Source* Author).

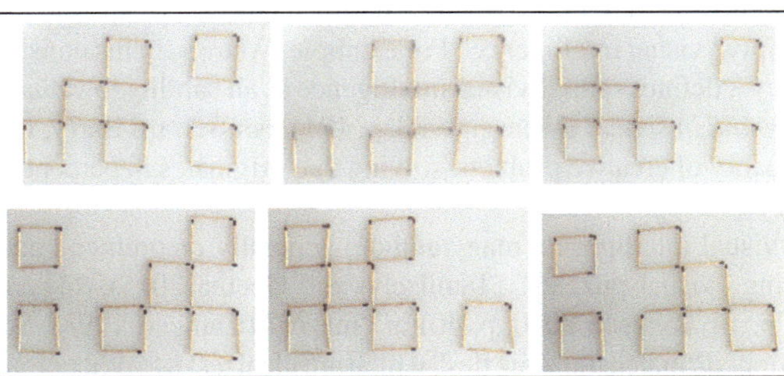

Fig. 11.2 Six different solutions of Tromholt's puzzle obtained by using visual intelligence (*Source* Author).

remain. Two solutions are possible". Czepa didn't repeat Tromholt's error and published two promised solutions (Czepa 1910, p.51). These solutions are given in Fig. 11.3.

Obviously in solving this puzzle Czepa had more visual intelligence than Tromholt. He found his second solution by moving matchsticks of all six squares to new positions.

Nevertheless, in Czepa's unrestricted puzzle formulation, *when squares shouldn't be of equal size*, there are more solutions to puzzle. Using Braun's strategy for making 6 unequal squares from 24 matches, one can find following class of fundamentally different solutions: two 2×2 squares (16 matchsticks used) and four 1×1 squares (8 matchsticks used). One these solutions is given in Fig. 11.4.

Using their visual intelligence, the readers will notice that in Czepa's unrestricted puzzle there are 16 additional solutions with unequal squares. First eight solutions are obtained by moving two matchsticks from upper 1×1 square into four different positions in left or right 2×2 square. Second eight solutions are obtained by moving two matchsticks from lower left 1×1 square into four different positions in left or right 2×2 square

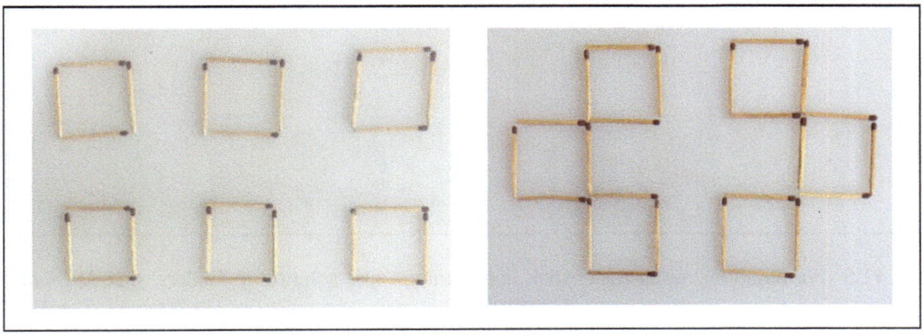

Fig. 11.3 Two different Czepa's solutions to Tromholt's puzzle (*Source* Author).

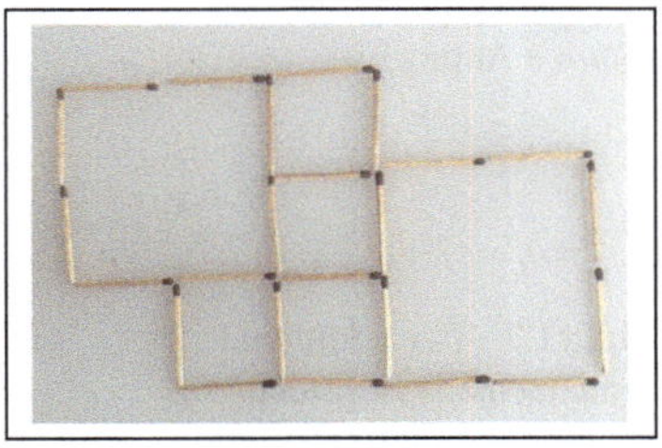

Fig. 11.4 A possible solution of Czepa's puzzle with unequal squares (*Source* Author).

11.3 Remove 12 Matchsticks, Leaving 3 Squares

The second example is a puzzle published by Lüders (1919, p. 46). Its formulation, initial matchstick configuration, and solution are given in Fig. 11.5.

After many years, Obermair (1977, p. 23, p. 96) and Brown (1978, p. 39, p. 177) reinvented Lüders' puzzle. As Lüders' solution was unknown to them, they publish a fundamentally different solution to this puzzle. It is given in Fig. 11.6.

Neither Lüders nor Obermair and Brown used their visual intelligence to find from their solutions two additional fundamentally different solutions. These solutions are given in Fig. 11.7.

Visual intelligence helps to notice that these solutions can be easily obtained from Lüders' solution by moving four matches of lower square to two different positions. Starting from Obermair's and Brown's solution, visually intelligent readers will see that the first solution can be obtained by moving matchsticks of middle square. The second solution is obtained by moving both middle and lower squares to new positions.

Remove twelve matches so that three squares remain.

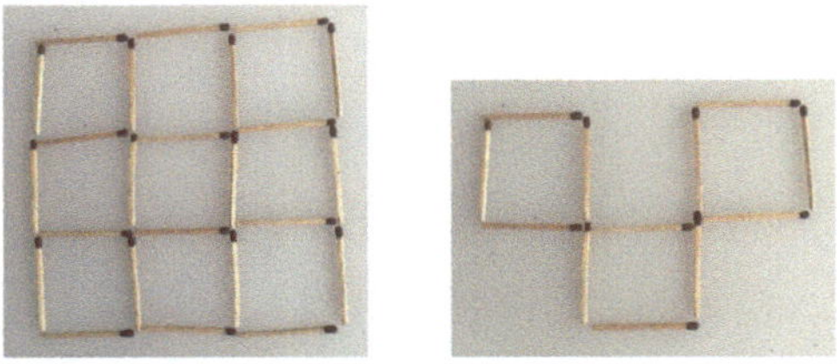

Fig. 11.5 Formulation, initial matchstick configuration and solution to Lüders' puzzle (*Source* Author).

Fig. 11.6 Solution of Lüders' puzzle found by Obermair and Brown (*Source* Author).

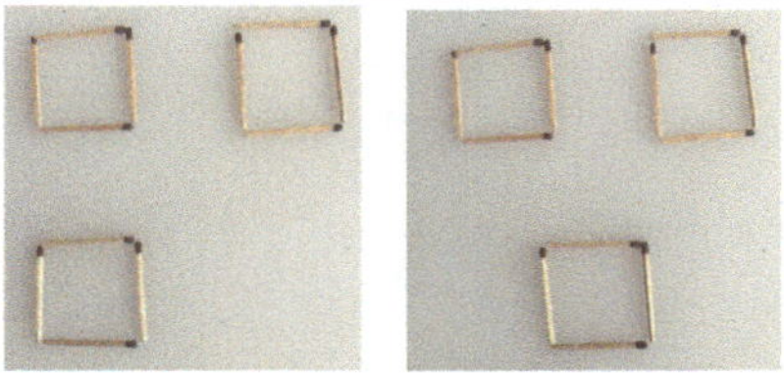

Fig. 11.7 Two additional fundamentally different solutions of Lüders' puzzle (*Source* Author).

11.4 Remove 5 Matchsticks to Make 4 Squares

Third example is a puzzle published by Obermair (1977, pp. 25–26, p. 114) and repeated by Hansell (Hansell 1981, p. 48, p. 89).

The formulation, initial matchstick configuration, and solution of this Obermair's puzzle is given in Fig. 11.8.

Insight into the solution, driven by visual intelligence, helps one to see that there are two classes of additional solutions which are fundamentally different.

In the first class, two matchsticks from the left 1 × 1 square in the lower row can be moved to form a new 1 × 1 square inside 2 × 2 square. In the second class, two matchsticks from the right 1 × 1 square in the lower row can be moved to form a new 1 × 1 square inside 2 × 2 square.

First additional solutions from both classes are given in Fig. 11.9.

Using combinatory thinking one can conclude: As in both classes the 1 × 1 square can have four different positions inside the 2 × 2 square, eight visually different solutions are possible.

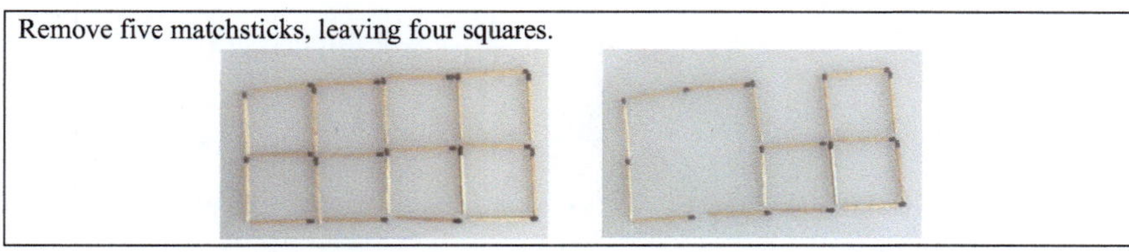

Fig. 11.8 Formulation, initial matchstick configuration, and solution to Obermair's puzzle (*Source* Author).

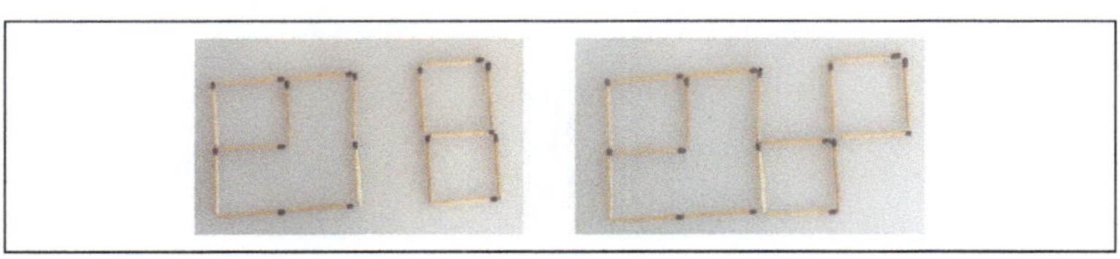

Fig. 11.9 First additional solutions to Obermair's puzzle from two classes.

11.5 Remove 9 Matchsticks to Make 4 Squares

The fourth and last example is again a puzzle published in English by Obermair. It was repeated by Hansell (1981, p. 49, p. 94) and Schuyt (1989, p. 63, p. 146).

Formulation, initial matchstick configuration, and solution are presented in Fig. 11.10.

Using visual intelligence, one can easily notice that there are three additional solutions. Two of them are fundamentally different and one is mirror image of the given solution.

Two matchsticks of the upper square can be moved to make first fundamentally different solution.

Three matchsticks from the upper square can be moved to make mirror image of the given solution.

Three matchsticks from the upper square can be moved to make the second fundamentally different solution.

These three visually different solutions are given in Fig. 11.11.

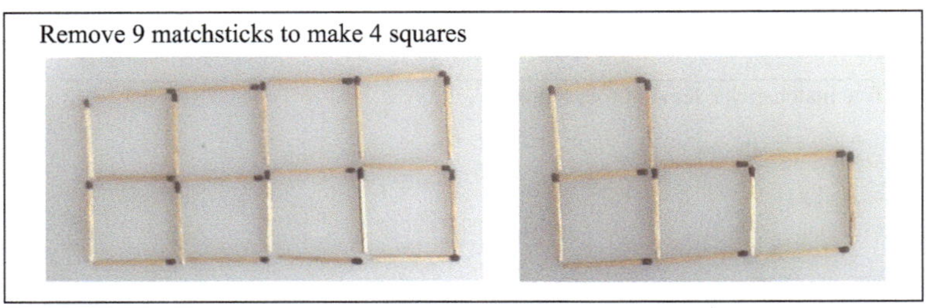

Fig. 11.10 Formulation, initial matchstick configuration, and solution to the second Obermair's puzzle (*Source* Author).

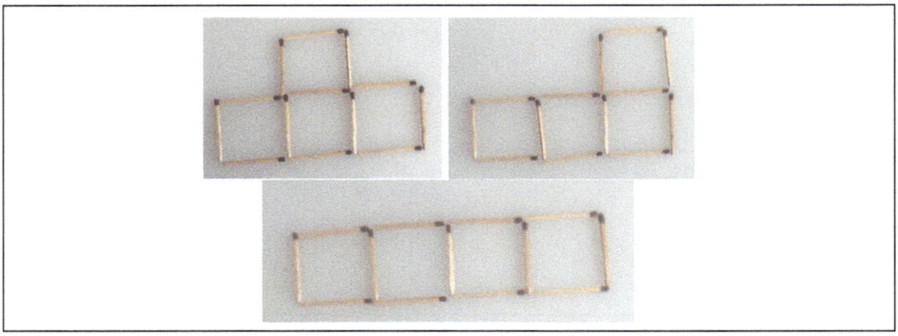

Fig. 11.11 Three visually different solutions to the second Obermair's puzzle (*Source* Author).

11.6 A Short Reflection

There is not any doubt that Sophus Tromholt, Alois Czepa, Gilbert Obermair, and many other book authors about matchstick puzzles were highly creative persons. It is then natural to ask a very difficult question:

Why did these authors in many multiple-solution matchstick puzzles publish only one solution?

I have two tentative answers.

The first answer is: Somehow, authors' creativity in posing matchstick puzzles was bigger than their creativity in solving these puzzles.

The second answer is: These authors in some multiple-solution matchstick puzzles didn't escape known cognitive bias "one puzzle—one solution" and failed to activate their visual intelligence.

Nevertheless, I expect that is possible teach how to use of visual intelligence to derive additional solutions to multiple-solution matchstick puzzle from the first found solution. It is my hope that future experiments will show that this expectation is correct one.

This hope is based on impressive experimental facts presented Chap. 10: Information about the number possible solution to multi-solution matchstick puzzle helped many puzzle solvers to find more solutions than famous books authors.

References

Brown, M. (Editor) (1978). *The great book of puzzles & perplexities.* Lyceum Books.

Cathcart, J. (1998). *The acorn principle. Know yourself - grow yourself. Discover, explore, and grow the seeds of your greatest potential.* St. Martin's Press.

Czepa, A. (1910). Streichholz-Spielereien: Kunststücke und Scherze nebst Auflösungen. Otto Maier Verlag.

Gardner, H. (1983). *Frames of mind. The theory of multiple intelligences.* Basic Books.

Hansell, S. (1981). *109 knifflige Streichholztricks.* Otto Maier Verlag.

Johnson, C. (2010). *It's all in your mind. Enhance your thinking skills for optimal success.* Hunter and Ryan Publishing.

Lüders, C. E. (1919). *Der Streichholzkünstler in der Westentasche.* Verlag G. Danner.

Obermair, G. (1977). Matchstick puzzles, tricks and games. Sterling Publishing Company.

Schuyt, M. (1989). *Phantastische Zündholzspiele.* DuMont Buchverlag

Seward Barry, A. M. (1997). *Visual intelligence. Perception, image, and manipulation in visual communication.* State University of New York Press.

Stine, J. M. (2002). *Super brain power. The 21-day programme to dramatically improve your intelligence.* Piatkus.

Tamilselvi, B., & Geetha, D. (2015). Efficacy in teaching through "multiple intelligence" instructional strategies. *Journal on School Educational Technology, 11*(2), 1–10.

Tromholt, S. (1889). *Streichholzspiele. Denksport und Kurzeil.* First edition. Otto Spamer Verlag.

Tromholt, S. (1890). *Streichholzspiele. Denksport und Kurzeil.* Fourth edition. Otto Spamer Verlag.

Tromholt, S. (1892). *Streichholzspiele. Denksport und Kurzeil.* Fifth edition. Otto Spamer Verlag.

Weaver, M. (1999. *Visual literacy. How to read and use information in graphic form.* Learning Express.

Chapter 12
Potential Utility of Matchstick Puzzles with Erroneous Solutions

Abstract This last chapter presents basic ideas for possible didactic use of matchstick puzzles with different types of erroneous solutions. As in other mathematical puzzles with erroneous solutions, one important possibility is to give puzzle solvers opportunities to practice and improve their skills of critical and creative thinking in solving unusual tasks related to matchstick puzzles. In the first case, they are asked to find the error, and, in the second case, their task is to propose a correct solution or to pose a new related puzzle. In thinking about such proposals, Braun's or Katona's strategies could be helpful. Proposed didactic sequences are structured for use in research, teaching, and home schooling.

12.1 Problems in Recreational Mathematics with Erroneous Solutions

Although recreational mathematics had an impressive bibliography 70 years ago (Schaaf 1955), recently appeared important books with synthetic and critical views about its history (Sesiano 2014; Singmaster 2021) and possible use in teaching problem solving (Averbach and Chein 2000). What is left out of focus is the presence of problems with erroneous solutions. In my view, there are, at least, three different class of such problems.

To the first class belong the problems with correct solving approach but with a careless numerical error in its realization that leads to an erroneous solution. One example is "drinking problem" published by Frisius (1540, p. 71):

> A drinker empties a jar of wine by himself in 20 days, while with the help of his wife, each drinking in proportion, they consume the wine in 14 days. How long would it take the wife alone to empty the entire jar?

Frisius assumed that the jar had 20 units of wine. Being so, the drinker consumed 14/20 or 7/10 of the jar. Instead of noticing that the wife consumed 6/20 or 3/10 of the jar, he said her part was 8/20 or 4/10 in 14 days. Such numerical error, in correct procedure with the rule of three, caused an erroneous answer of 35 days.

J. Slisko, *Fostering Cognitive Mathematics Skills with Matchstick Puzzles*, Mathematics in Mind, https://doi.org/10.1007/978-3-032-14561-1_12

To the second class belong the problems with contextually incorrect solving approach. The most famous example is Fibonacci's problem dealing with an "ascending and descending lion" (Sigler 2002, p. 273):

> A certain lion is in a certain pit, the depth of which is 50 palms, and he ascends daily 1/7 of a palm, and descends 1/9. It is sought in how many days will he leave the pit.

Fibonacci rightly found that in 63 days the lion would ascend 2 palms. Based on the rule of three, his incorrect solution was 1,575 days. With contextually correct approach, in which the lion gets to the top of the pit not by descending but by ascending, the solution is approximately 1,572 days.

In the third class are the recreational problems with a "*didactic error*". Namely, the readers are given a correct solution without the solving approach used to get it. A good example is Dudeney's puzzle "Sir Edwyn de Tudor" (Dudeny 1917, p. 17):

> Sir Edwin de Tudor (is) going to rescue his lady-love, the fair Isabella, who was held a captive by a neighboring wicked baron. Sir Edwyn calculated that if he rode fifteen miles an hour, he would arrive at the castle an hour too soon, while if he rode ten miles an hour, he would get there just an hour too late. Now, it was the first importance that he should arrive at the exact time appointed, in order that the rescue that he had planned should be a success, and the time of the tryst was five o'clock, when the captive lady would be taking he afternoon tea. The puzzle is to discover exactly how far Sir Edwyn de Tudor had to ride.

The solution given to readers was (Dudney 1917, 175):

> The distance must have been sixty miles. If Sir Edwyn left at noon and rode 15 miles an hour, he would arrive at four o'clock – an hour too soon. If he rode 10 miles an hour, he would arrive at six o'clock, he would reach the castle of the wicked baron exactly at five o'clock – the time appointed.

Obviously, the readers are obliged to find their own solving paths leading to this solution. In a similar sense, all matchstick puzzles share the same "*didactic error*" because the readers aren't informed about possibilities of strategic thinking in solving geometric matchstick puzzles (Braun 1876; Katona 1940).

In two small-scale studies, students were able to recognize and correct a careless numerical error in mathematics textbook of Gaspar Nicolas (Costa Clain and Slisko 2025) and perform better than Fibonacci in "lion problem" (Slisko 2020). In addition, they acknowledged that knowing the errors, mathematicians made in past, helps them to change their negative feelings on own mistakes in mathematics learning. In other words, they became better prepared to learn from own errors (Tulisa et al. 2016).

In what follows, I present some basic ideas for possible didactic use of matchstick puzzles with different types of erroneous solutions that might cause the same positive effects in students and general readers.

12.2 Solution to a Matchstick Puzzle with a "Free" Matchstick

Louis Mittenzwey, a recognized mathematics teacher and school director, published a matchstick puzzle with erroneous solution (Mittenzwey 1880, p. 37, p. 90). Strangely, the error was waiting 125 years to corrected by Heinrich Hemme in recently published nineth edition of the book (Mittenzwey, 2025, p. 76, p. 181, p. 237)!

The formulation, initial matchstick formulation, and "solution" to the puzzle are given in Fig. 12.1.

Critical thinking

Why the given solution is wrong?

Expected answer: Four squares are formed, but one matchstick is unused.

Creative thinking

What are two correct solutions with four squares in which unused matchstick will became side of a square after moving other matchsticks?

Expected solutions: In the first solution one matchstick is moved, and in the second solution two matchsticks are moved. Expected solutions are given in the Fig. 12.2.

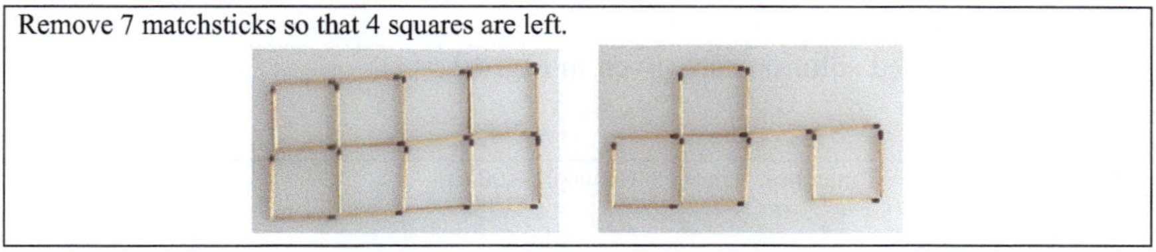

Fig. 12.1 Formulation and solution of Mittenzwey's puzzle (*Source* Author)

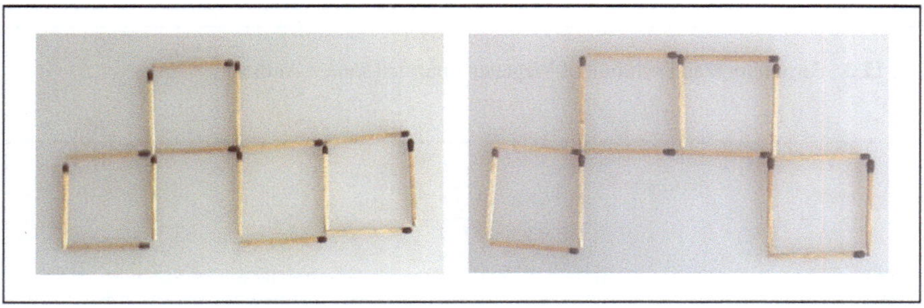

Fig. 12.2 Two correct expected solutions to Mittenzwey's puzzle (*Source* Author)

12.3 Solutions with a Missing Square

Wegener (1930, p. 135, p. 15) published a matchstick puzzle in which the solution had one square less. The formulation, initial matchstick formulation, and solution of Wegener's puzzle are given in Fig. 12.3.

Critical thinking

Why the given "solution" is wrong?
Expected answer: Instead of three squares, two squares are formed.

For which puzzle would this solution be correct?
Expected answer: Remove 6 to make 2 squares.

Creative thinking

What are possible different correct solutions for the Wagener's puzzle: Remove 6 to make 3 squares? To get an idea use Braun's strategy.
Expected answers: After removing 6 matchsticks, 11 matchsticks are left for making 3 squares.

1. Supposing that all squares are equal, the structure of solutions is:

   Two connected 1 × 1 squares (made from 7 matchsticks) and one disconnected 1 × 1 square made from 4 matchsticks.
   Three of expected solutions are given in Fig. 12.4.

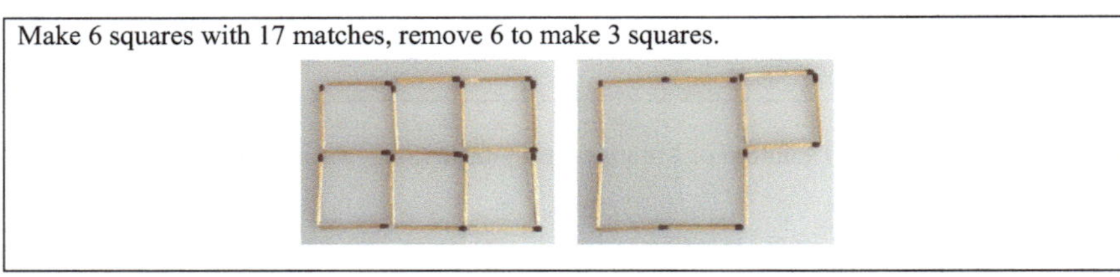

Fig. 12.3 Formulation and solution of Wegener's puzzle (*Source* Author)

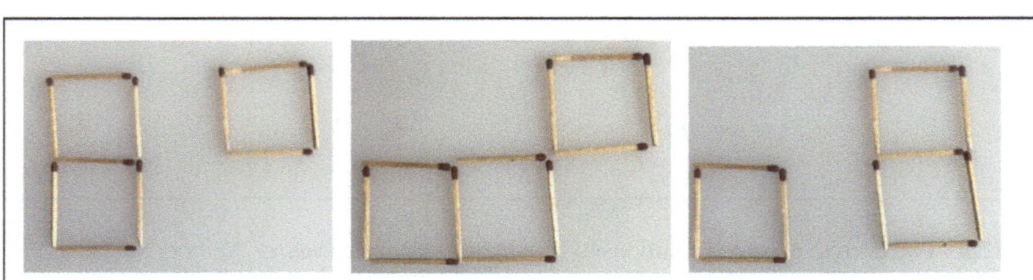

Fig. 12.4 Three expected solutions of Wegener's puzzle if remaining squares are of equal size (*Source* Author)

2. Supposing that squares are unequal, the structure of solutions is:

One 2 × 2 square made from 8 matchsticks and inside it two 1 × 1 squares made from 3 matchsticks.

Three of expected solutions are given in Fig. 12.5.

Bogart published a puzzle whose solution had a missing square, too (Bogart 2023, p. 28, p. 63). Its formulation, initial matchstick configuration, and solution are given in Fig. 12.6.

Critical thinking

Why the given solution is wrong?

Expected answer: Five squares are formed instead of required six squares.

Creative thinking

What are 3 of 16 visually different correct solutions? Use Katona's strategy to get an idea how to transform incorrect into correct solution.

Expected use of Katona's strategy and solutions: To increase the number of squares by one, it is necessary to eliminate one old 1 × 1 square and build from its matchsticks two new 1 × 1 squares inside 2 × 2 square. Three correct solutions to Bogart's puzzle are given in Fig. 12.7.

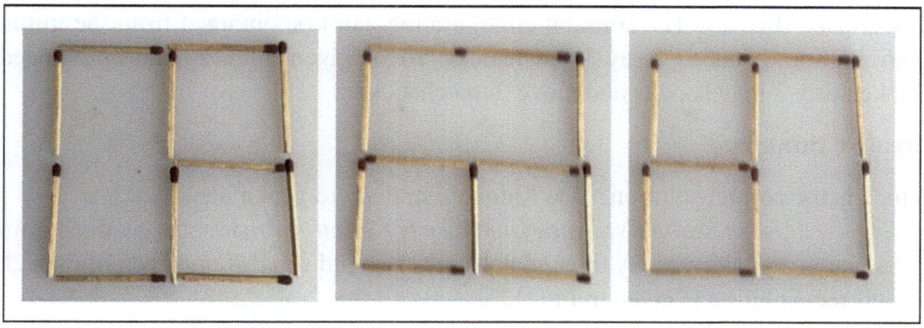

Fig. 12.5 Three expected solutions of Wegener's puzzle if remaining squares are of unequal size (*Source* Author)

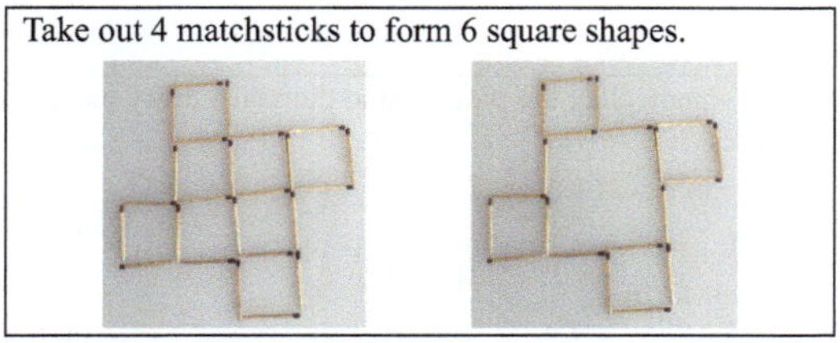

Fig. 12.6 Formulation and solution of Bogart's puzzle (*Source* Author)

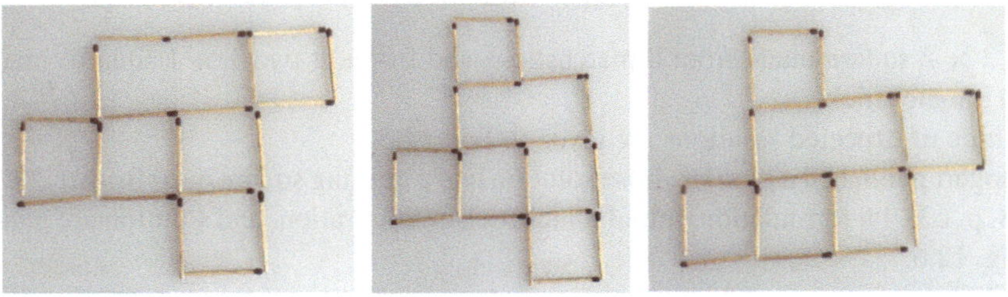

Fig. 12.7 Three correct solutions of Bogart's puzzle (*Source* Author)

12.4 A Puzzle with an Inverted Solution

An anonymous author published in the journal "*American Agriculturist*" a puzzle with inverted solution (Anonymous 1911a, b). Formulation, initial matchstick formulation, and solution to that puzzle are given in Fig. 12.8. Instead of original word "toothpicks", the word "matchstick" is used.

Critical thinking

Why the solution is wrong?

Expected Answer: The solution configuration can't be obtained from the initial configuration by taking away 3 matchsticks. To get given solution, one would need to take away 3 matchsticks and move 2 matchsticks.

Creative thinking

Which is the correct solution? Use Katona's strategy to get an idea.

Expected Use of Katona's Strategy and Correct Solution Answer: To get three from five squares, by taking away three matchsticks two old squares must be eliminated. The correct solution is given in Fig. 12.9.

How could you obtain the correct solution from the wrong one (Fig. 12.8) without moving a single matchstick?

Expected Answer: Using a vertical mirror. Mirror image of wrong solution would be correct solution (Fig. 12.10).

The puzzle is to make five squares like the picture, and then take away just three matchsticks, and leave three complete squares.

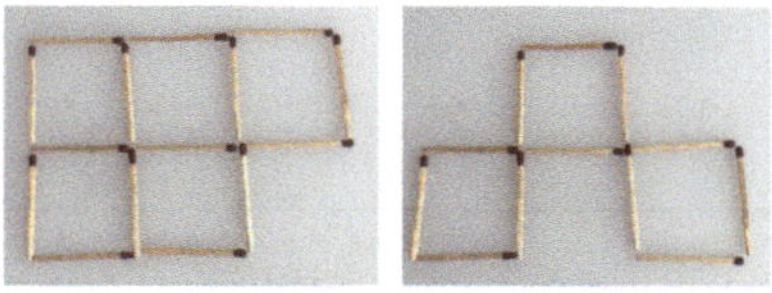

Fig. 12.8 Formulation, initial matchstick configuration and solution to the puzzle in the journal "*American Agriculturist*" (*Source* Author)

Fig. 12.9 Correct solution to the puzzle in the journal American Agriculturist (*Source* Author)

Fig. 12.10 Correct solution as a mirror image of wrong solution (*Source* Author)

12.5 Careless Use of Words "Remove" and "Move"

Costello published a puzzle in whose formulation the word "move" is used instead of the word "remove" (Castello 1996, pp. 138–139, p. 152). Formulation, initial matchstick configuration, and solution to Costello's puzzle are given in Fig. 12.11.

Critical thinking

Why Costello's solution is wrong?

Expected answer: Two matchsticks weren't moved but removed.

To which puzzle Costello's solution would be correct?

Here is one of the classic match problems. The object is to move two matches to make two squares instead of four.

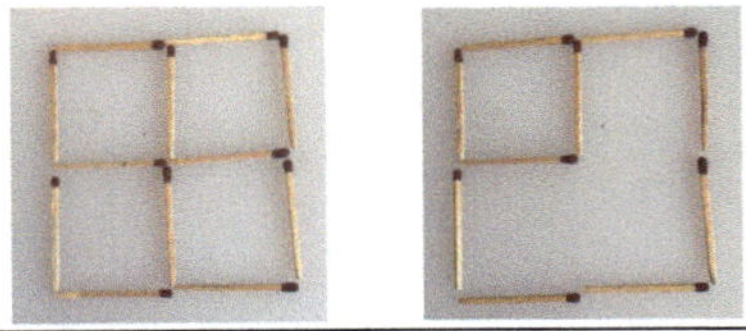

Fig. 12.11 Formulation, initial matchstick configuration, and solution to Costello's puzzle (*Source* Author)

Expected answer: Remove two matches to make to make two squares instead of four.

Creative thinking

It is impossible to make 2 squares by moving 2 matches in Costello's initial configuration. How many matchsticks should be moved to make 2 squares? Use Braun's strategy to get an idea.

Expected use of Braun's strategy and corresponding solutions: There are 12 matchsticks which should be reorganized to make two squares. The only possibility is to make one 2 × 2 square (with 8 matches) and one 1 × 1 square (with 4 matches).

This can be done in two ways getting four solutions. The first two solutions are obtained by moving 4 inside matches to make one 1 × 1 square, either inside or outside the 2 × 2 square. The second two solutions are obtained by moving 6 matches to leave one existing 1 × 1 and making a new 2 × 2 square. The new 2 × 2 square can be either unconnected or connected.

Four expected solutions are given in Fig. 12.12.

Anonymous author from Regionalia Verlag published a puzzle in whose formulation the word "remove" is used instead of the word "move" (Anonymouos 2014, p. 38, p. 102). Its formulation, initial matchstick configuration and solution are given in Fig. 12.13.

Critical thinking

Why the given solution is wrong?

Expected Answer: Two matches weren't removed but moved.

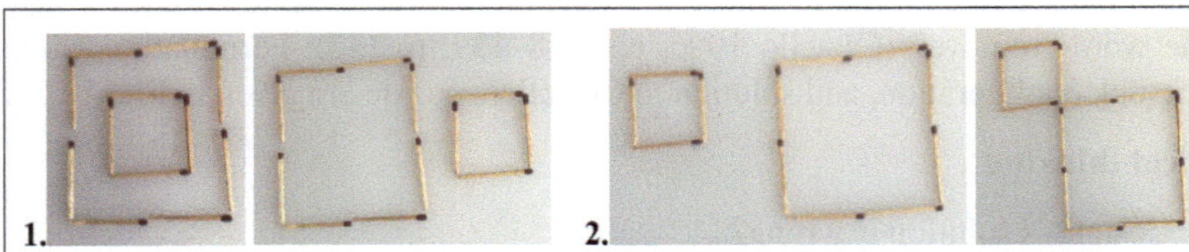

Fig. 12.12 Four correct solutions to Costello's puzzle with moved matches (*Source* Author)

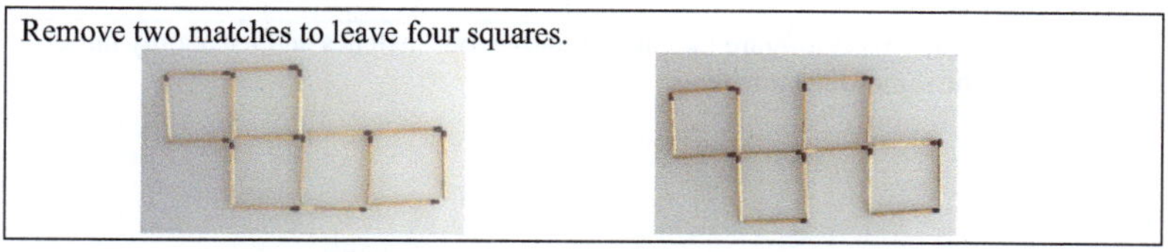

Fig. 12.13 Formulation, initial matchstick formulation, and solution to the puzzled published by anonymous author from Regionalia Verlag (*Source* Author)

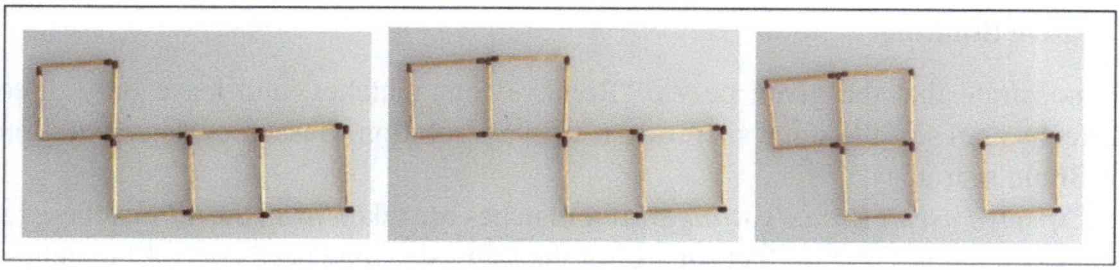

Fig. 12.14 Three correct solutions to the puzzle published by the anonymous author from Regionalia Verlag (*Source* Author)

For which puzzle would be given solution correct?

Expected Answer: Move two matches to leave four squares.

Creative thinking

Which are three correct solutions for the given puzzle: Remove two matches to leave four squares? Use Katona's strategy to get an idea.

Expected Use of Katona's Strategy and Correct Solution: To get four from five squares, one old square must be eliminated by removing two matchsticks. Three resulting correct solutions are given in Fig. 12.14.

12.6 A Puzzle Without a Solution

Grose published a puzzle which had a wrong solution and, additionally, was unsolvable as formulated (Grose 1928, p. 114, p. 117). Its formulation, initial matchstick configuration, and solution are given in Fig. 12.15.

Critical thinking

Why the given solution is wrong?

Expected Answer: There are four squares not three squares.

For which puzzle the given solution would be correct?

Expected Answer: Remove four matches and leave four perfect squares.

With seventeen matches make a figure like this. Remove four matches and leave only three perfect squares of the same size remaining.

Fig. 12.15 Formulation and solution to Grose's puzzle (*Source* Author)

Creative thinking

Demonstrate that the given puzzle "Remove four matches, and leave only three perfect squares of the same size" doesn't have a solution. Base your demonstration on Braun's strategy.

Expected use of Braun's strategy: If 4 matches are taken from the 17 matches, 13 matches remain. For 3 perfect squares of the same size, one needs only 12 matches. So, this puzzle has no solution.

Formulate your puzzles with three remaining equal squares that are solvable. Present two visually different solutions for each of your puzzle.

Expected puzzles and solutions

Puzzle 1. Remove five matches to get three equal squares. Its two visually different solutions are given in Fig. 12.16.

Puzzle 2. Remove six matches to get three equal squares. Its two visually and fundamentally different solutions are given in Fig. 12.17.

Fig. 12.16 Two visually different solutions to the Puzzle 1 (*Source* Author)

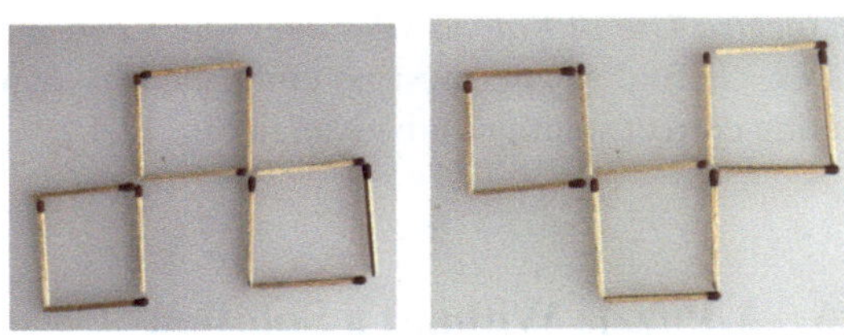

Fig. 12.17 Two visually different solutions to the Puzzle 2 (*Source* Author)

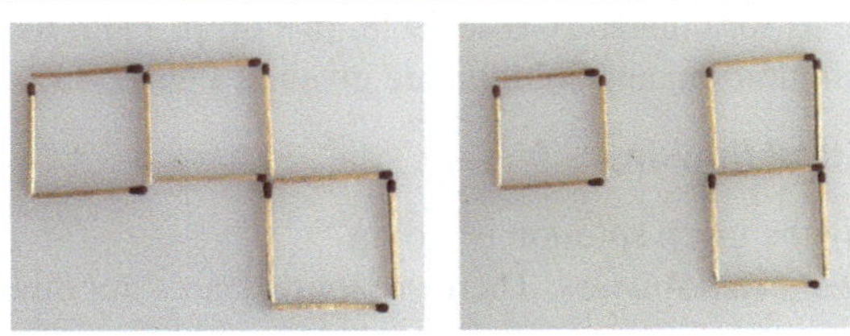

References

Anonymous (1911a). Another toothpick puzzle. *American Agriculturist, 88*(20), 495.

Anonymous (1911b). Answer to toothpick puzzle. *American Agriculturist, 88*(21), 519.

Anonymous (2014). *Das fabelhafte Büchlein der besten Streichholzspiele.* Regionalia Verlag.

Averbach, B. & Chein, O. (2000). *Problem solving through recreational mathematics.* Dover Publications

Bogart, S. (2023). *99 matchstick puzzles. Puzzles to improve your math skills.* Coppell.

Braun, F. (1876). *Der junge Mathematiker und Naturforscher.* Verlag von Otto Spamer.

Costa Clain, T. & Slisko, J. (2025). Using the history of mathematics as a methodological strategy to understand Gaspar Nicolas' "errors". *International Electronic Journal of Mathematics Education, 20*(4), em0846.

Costello, M. J. (1996). *The greatest puzzles of all time.* Dover Publication.

Dudeney, H. E. (1917). *Amusements in mathematics.* Thomas Nelson and Sons.

Frisius, G. (1540). *Arithmeticae praticae methodus facilis.* Gregorio Bontio.

Grose, H. B. (Editor) (1928). *Our wonder world. A library of knowledge. Volume Ten. The quiz book.* Geo L. Shuman and Company.

Katona, G. (1940). *Organizing and memorizing. Studies in the psychology of learning and teaching.* Columbia University Press.

Mittenzwey, L. (1880). *Mathematische Kurzweil.* First edition. Verlag Julius Klinkhardt.

Mittenzwey, L. (2025). *Mathematische Kurzweil.* Nineth edition. Anaconda.

Sesiano, J. (2014). *Récréations mathématiques au Moyen Âge.* EPFL Press.

Sigler, L. D. (2002). *Fibonacci's Liber Abaci. A translation into modern English of Leonardo Pisano's book of calculation.* Springer.

Singmaster, D. (2021). *Adventures in recreational mathematics.* 2 volumes. World Scientific.

Schaaf, W. L. (1955). *A Bibliography of Recreational Mathematics,* 4 volumes. National Council of Teachers of Mathematics.

Slisko, J. (2020). What students can learn from Fibonacci´ s error in solving "The lion in the pit" problem. *Góndola, enseñanza y aprendizaje de las ciencias, 15*(2), 216–238.

Tulisa, M., Steuera, G. & Dresela, M. (2016). Learning from errors: A model of individual processes. *Frontline Learning Research, 4*(2), 12–26.

Wegener, A. B. (1930). *Play games and other play activities.* The Abingdon Press.

Conclusions

I hope that this book provided initial, research-based answers to two important questions:

> What are common inadequacies in actual treatments of matchstick or toothpick puzzles in books for general readers and mathematics textbooks?
>
> How should matchstick or toothpick puzzles be pedagogically treated in formal and informal education?

The answers to the first question are:

Book and textbook authors commonly publish only one solution to multiple-solution matchstick puzzles, revealing cognitive bias "one puzzle—one solution". Doing so, they foster the same false belief in readers and students. This false belief is a main obstacle to creative thinking in matchstick puzzles solving.

Book and textbook authors neither present nor discuss two types of different solutions to multiple-solution matchstick puzzles:

(1) visually different solutions between which exist rotation and/or reflection relationships;
(2) fundamentally different solutions between which neither rotation nor reflection relationships exist.

Some book and textbook authors show cognitive bias "formulation change doesn't matter", publishing for unrestricted formulation only single solution to restricted solution. Commonly, unrestricted formulation makes possible additional (visually or fundamentally different) solutions with unequal squares or triangles.

The answers to the second question are:

Puzzle solvers in formal and informal education should be initially informed by various examples about what are visually different and fundamentally different solutions to matchstick puzzles.

Puzzle solvers should be informed about the number of visually or fundamentally different solutions in every matchstick puzzle they deal with and should face enough

J. Slisko, *Fostering Cognitive Mathematics Skills with Matchstick Puzzles*, Mathematics in Mind, https://doi.org/10.1007/978-3-032-14561-1

tasks in which they would be able to find these solutions and verify their supposed features (existence or absence of rotation and reflection relationships).

Puzzle solvers should be informed by various examples about Braun's and Katona's strategies for finding general structure of sought solutions. They should have enough tasks in which they can practice and improve these solving strategies.

Puzzle solvers should be also guided to a higher creativity level at which they would be able to formulate and solve their own matchstick puzzles. The best proposed puzzles would be those having different solutions.

After carried out a few initial, small-scale investigations with very positive results (information about the number of solutions stimulates creativity of puzzle solvers!) presented in Chap. 10, I feel that I'm in position to propose some future experimental research lines to get more theoretical knowledge about the process of matchstick puzzles solving and related pedagogies of teaching those puzzles to students of mathematics.

The objectives of these research lines might be:

provide more experimental results that adequate didactic teaching forms, similar to the pedagogical approach used by Katona, improve performance of puzzle solvers, both in puzzle solving as in puzzle posing;

demonstrate, with bigger and diverse groups of participants, that information about the number of possible different solutions in multiple-solution matchstick puzzles increments the number of solutions puzzle solvers would be able to find;

explore experimentally which verbal hints after formulation help matchstick puzzles solvers find all different solutions;

explore experimentally what are changes in self-estimation of puzzle solvers when they find out that they are able to find those puzzle solutions which were missed by famous book authors;

explore experimentally how known measurable cognitive skills, like logical thinking, visual intelligence and metacognition, are related with success or failure levels in matchstick puzzle solving and posing, and in finding and correcting erroneous solutions;

design and implement intelligent computer-based environments that permit experimental exploration of used time and number of successful and unsuccessful attempts in solving arithmetic and geometric puzzles with moving and removing virtual matchsticks in initial configurations.

The results of these researches would be a base to understand better how solving matchstick puzzles foster development of cognitive mathematics skills in formal and informal education.

Appendix

Written by Heinrich Hemme

Braun's Puzzle "Remove 5 Matchsticks to Get 5 Triangles"

Formulation and initial matchstick configuration of Braun's puzzles are given in Fig. A1.

As nothing was said about the sizes of the triangles, the solutions can contain triangles with side lengths of one matchstick (1), of two matchsticks (2), and of three matchsticks (3). Since it consists of five triangles in total, the possible combinations of triangles are 32221, 32211, 32111, 31111, 22111, 21111, and 11111.

It is easy to verify that there is no solution with triangle 3. The case 22211 must also contain triangle 3 and therefore does not yield a solution.

By systematically checking, one finds five solutions for 22111, nine for 21111, and seven for 11111.

From the figure below remove 5 matchsticks to get 5 equilateral triangles.

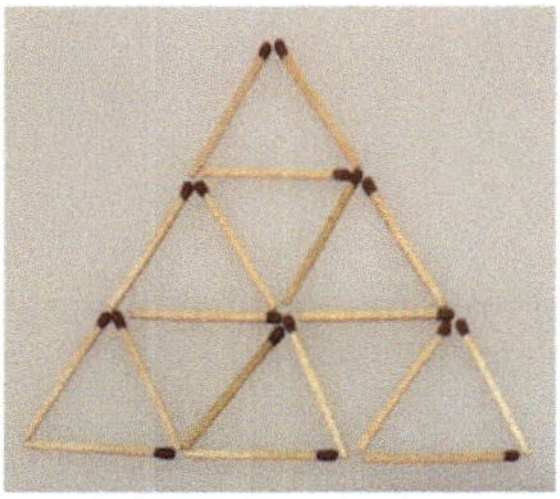

Fig. A1 Formulation and initial matchstick configuration of Braun's puzzle (*Source* Author)

J. Slisko, *Fostering Cognitive Mathematics Skills with Matchstick Puzzles*, Mathematics in Mind, https://doi.org/10.1007/978-3-032-14561-1

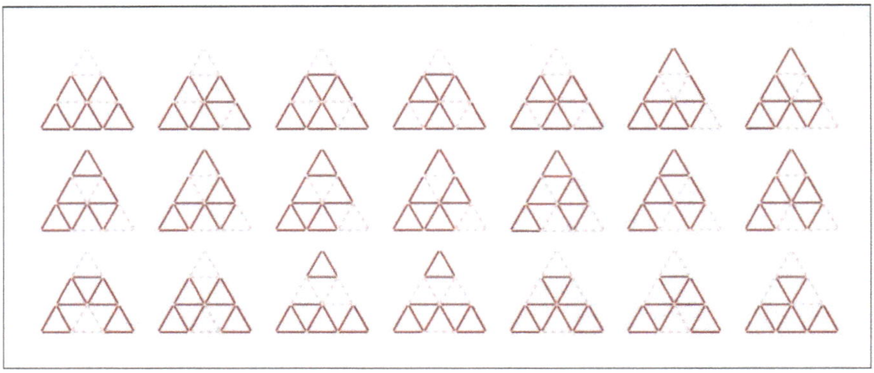

Fig. A2 Twenty-one solutions to Braun's puzzle (*Source* Heinrich Hemme)

Patterns that merge into one another through rotations or reflections are considered equal.

Twenty-one solutions are presented in Fig. A2.

Tromholt's Puzzle "Remove 8 Matchsticks to Get 5 Equal Squares"

Formulation and initial matchstick configuration of Tromholt's puzzle are given in Fig. A3.

Twelve pentominos are possible candidates for the solutions to Tromholt's puzzle with five remaining equal squares (Fig. A4).

Each of the I-, L-, N-, and Y-pentominos have four squares in a row. They are therefore ruled out.

The P-pentomino contains a block of four squares. Therefore, nine matches must be removed.

The X-pentomino is Tromholt's solution, and the six pentominos V, W, Z, F, T, and U are additional solutions.

Fig. A3 Formulation and initial matchstick configuration of Tromholt's puzzle (*Source* Author)

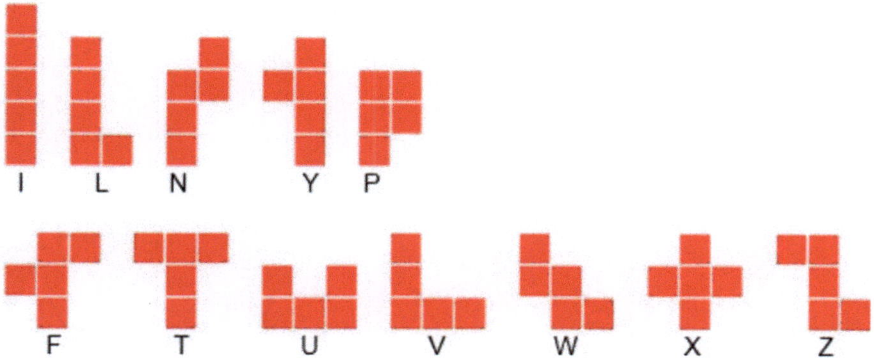

Fig. A4 Twelve pentominos with common associated letters (*Source* Heinrich Hemme)

MIX
Papier aus verantwortungsvollen Quellen
Paper from responsible sources
FSC® C105338

If you have any concerns about our products,
you can contact us on
ProductSafety@springernature.com

In case Publisher is established outside the EU,
the EU authorized representative is:
Springer Nature Customer Service Center GmbH
Europaplatz 3, 69115 Heidelberg, Germany

Printed by Libri Plureos GmbH
in Hamburg, Germany